AF711476

AIRCRAFT WOODWORK

DRAKE'S AIRCRAFT MECHANIC SERIES

AIRCRAFT WOODWORK

AIRCRAFT WELDING

AIRCRAFT SHEET METAL

AIRCRAFT ENGINES

AIRCRAFT ELECTRICAL SYSTEMS,
HYDRAULIC SYSTEMS,
AND INSTRUMENTS

AIRCRAFT MAINTENANCE AND SERVICE

AIRCRAFT ENGINE MAINTENANCE AND SERVICE

AIRCRAFT WOODWORK

by COLONEL ROLLEN H. DRAKE, B. S., M. A.

RESEARCH ENGINEER, PILOT AND GROUND SCHOOL INSTRUCTOR FORMERLY: INSTRUCTOR, VOCATIONAL AND AVIATION SUBJECTS, LOS ANGELES CITY SCHOOLS; CHIEF, AIR AGENCY UNIT AND CIVILIAN PILOT TRAINING SPECIALIST, CIVIL AERONAUTICS ADMINISTRATION; SUPPLY SPECIALIST, OFFICE OF THE QUARTERMASTER GENERAL

Originally published in 1946

ISBN: 9781940001319

The Aviation Collection
by
Sportsman's Vintage Press
2015

This book is dedicated to

MY BELOVED WIFE

without whose inspiration and help

this book could not have been written

PREFACE

WOOD has been used extensively in the fabrication and repair of aircraft since man first attempted to conquer the air. This book has been planned to fill an ever-increasing need for a single volume, written in non-technical language, covering all phases of aircraft woodwork. It is suitable for use as a mechanic's manual or as a textbook in the classroom and shop.

The text contains considerable information on the characteristics, identification, and uses of aircraft woods. The methods and materials used in the fabrication of wood aircraft structures have been described in detail. The practical problems cover the important phases in the fabrication and repair of wood aircraft structures.

This text has been written for use in colleges having vocational training departments; junior colleges in their vocational and terminal courses; high schools; trade and vocational schools; and mechanics' schools. The mechanic in the field should find all of the information necessary for repairs and service to the wood parts of an airplane. The text is designed to cover approximately one year's work in a standard vocational curriculum. It is anticipated that this book will be of great value in the rehabilitation training programs. The casual reader who is interested in this subject will find much of value.

It is the opinion of many experts in the field of aviation that wood will be used extensively in aircraft manufacturing for many years. There is a continued and increasing demand for the aircraft woodworker.

The author wishes to express his grateful appreciation to the following who have so kindly furnished material which has been of great assistance in the preparation of this book: the United States Office of Education; Civil Aeronautics Administration; Forest Products Laboratory; Universal Moulded Products Corporation; the Departments of Education of the various states, particularly New York, Pennsylvania, Utah and Virginia; Douglas Aircraft Company; Piper Aircraft Corporation; Curtiss-Wright Corporation; and Timm Aircraft Corporation.

The author wishes to thank his many friends who have so generously contributed their advice and assistance. He also wishes particularly to express his gratitude to Earle R. Hough for his excellent work in the preparation of many of the drawings in this book, and to Mildred Pickrel for her untiring patience and cooperation in the preparation of the manuscript.

The frontispiece is a photograph of Sitka Spruce Forest, reproduced by courtesy of American Forest Products Industries.

R. H. D.

TABLE OF CONTENTS

AIRCRAFT WOODWORK

INTRODUCTION

Imagine primitive man's desire to travel when he first lifted his eyes from the soil and gazed upon the horizon. As he plodded slowly toward the ever distant horizon, he must have seen the apparently effortless flight of the birds. Although he was but a ground plodder, he must have, even then, felt the desire to fly. Mythology and earliest history are filled with man's desire to imitate the birds in flight. The winged horse of the Greeks and the winged shoes of Mercury were but indications of this desire.

The use of wood was man's first step toward civilization. Wood has been the best friend of man through the ages. It not only kept him warm, but furnished his first shelter after he left the caves. Later he found that a piece of wood in his hands became a weapon, a club with which to help him gain his daily food. He probably first made use of wood as a tool in this manner. Then he used it as a lever to pry up logs or rocks in the search of food. Later he found that he could throw a stick of wood and by accident found that when it traveled end first it caused more serious injury to the game. He sharpened one end of a stick and learned to use it as a spear. Primitive man also found that by partially charring the sharpened stick the point became harder and more effective as a weapon of self-defense and self-preservation.

Man undoubtedly used a floating log or tree as the first means of travel other than by his own effort — first floating with the current and later learning to paddle the log and make it take him where he wished.

The Age of Wood lasted for thousands of years. The first use of stone tools was to work wood. As man learned to make better tools, his skill in the use of them improved, and he built a more highly finished home. The next step was probably the making of furniture — crude at first, and then the more elaborate, beautiful, and artistic furniture of the middle ages. Many household utensils were first made from wood. The wood carver and the cabinetmaker were the highest types of tradesmen.

Much woodworking today is done by means of machinery, but the demand for the skilled worker in wood, using hand tools, has kept pace with civilization.

The author had occasion to visit one of the largest aircraft factories in the country a few years ago. When we came to the section where wings were constructed and fabricated, the guide called attention to the fact that every man working in this section was past fifty-five years of age, and each of them had learned his trade as a cabinetmaker in one of the European countries. The guide stated that, due to the general use of machines in the working of wood in the United States, it was impossible to find enough men skilled in the use of hand tools to perform the exacting operations demanded in the fabrication of airplanes. During the last few years there has been a continuing demand for skilled woodworkers.

This book on aircraft woodwork is planned to assist you to become a

Fig. 1. Sitka spruce (aircraft wood) on its way to the band saw. (Courtesy American Forest Products Industries, Inc.)

Fig. 2. A log being cut into lumber as it moves past the band saw. (Courtesy American Forest Products Industries, Inc.)

Fig. 3. Aircraft woodworker in a modern factory. (Courtesy Timm Aircraft Corporation)

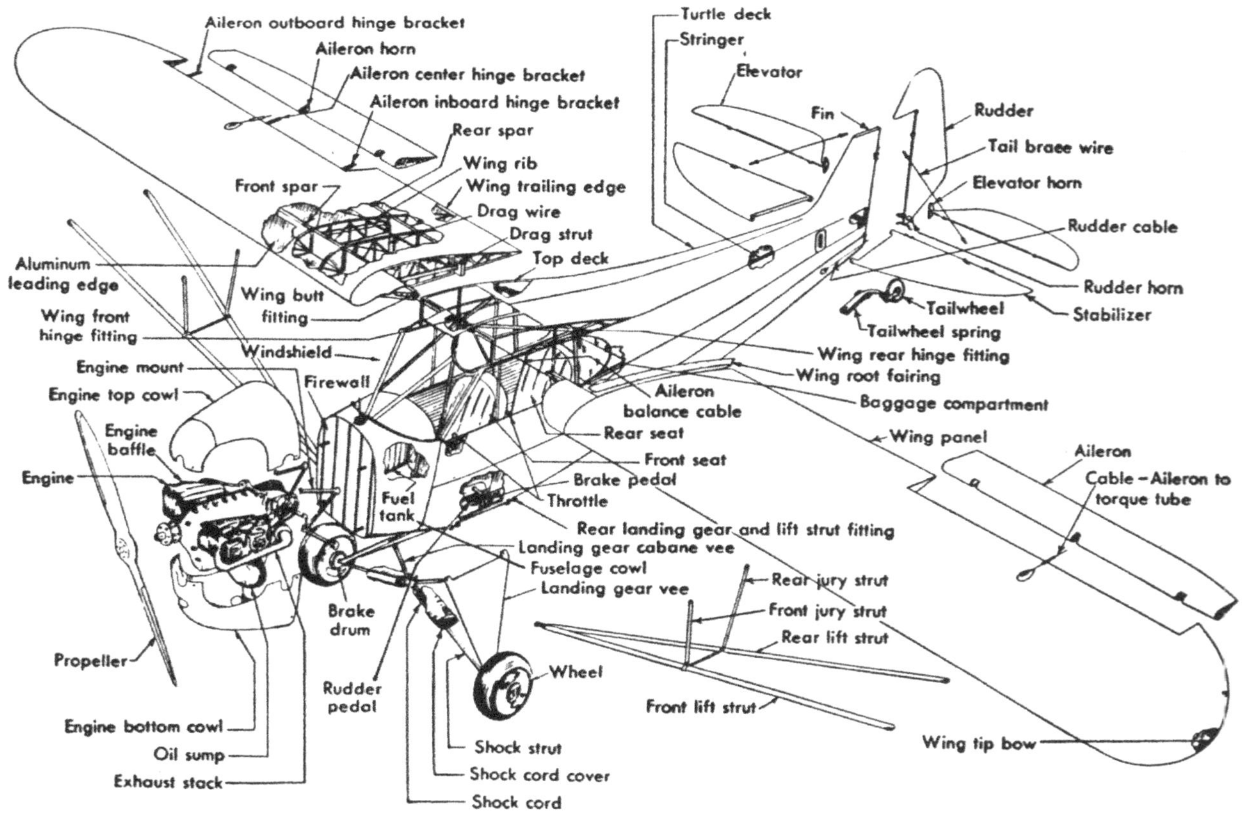

Fig. 4. Visual view of a modern light airplane. (Courtesy Piper Aircraft Corporation)

skilled worker in wood, whether you are sixteen or sixty. The skills developed will apply to any of the industries where wood is used. Wood is friendly to the worker who understands his tools. It is plentiful and will continue to be used in all of the industries for many years to come.

Aircraft Nomenclature. Before beginning actual work on wood in aircraft construction, it is necessary to become familiar with the various

Fig. 5. A modern light airplane with a two-place cabin. (Courtesy Piper Aircraft Corporation)

parts of the airplane structure. The main divisions of an airplane are: (1) the body of the airplane, called the fuselage; (2) the airfoils, or wings, which, through their reaction to the air, support the weight of the airplane in flight; (3) the tail group, called the empennage, consisting of the rudder, elevators, vertical fin, and horizontal stabilizer; (4) the landing gear, which includes the wheels, tail wheel, shock absorbers, and the necessary bracing; and (5) the engine mount, which is the main support for the engine, and the necessary cowling to enclose the engine.

The fuselage of all airplanes is similar in design and location. The main difference is in size and the use for which the airplane is designed. A small, light aircraft has a fuselage arranged to seat one or two passengers, including the pilot. With a few pounds of baggage, they make up the entire payload of the airplane. The large transport planes have rows of seats arranged to carry many passengers, with room in the front of the fuselage for baggage, radio equipment, etc. In this type of airplane, the pilot and copilot are far forward in the fuselage and in a separate compartment from the passengers. The fuselage of cargo-carrying airplanes has large spaces in which cargo may be stored.

Fig. 6. A modern transport airplane showing the pilots' compartment in the nose. (Courtesy Douglas Aircraft Company, Inc.)

In light airplanes, the fuselage may be made of welded steel tubing and covered with fabric; a wood framework covered with light plywood; a combination of metal and plywood; or wood and fabric. The types of wood construction used in the various kinds of airplanes will be taken up and explained more fully later on in this book.

The wings are much the same for every type of airplane. The method of construction varies much more widely than the shape. In many military aircraft, large transport, and cargo airplanes, the wings have no external support and are made almost entirely of metal. This type of construction is called "cantilever."

In the light airplanes more wood than metal is generally used, and the wings have some outside support, such as wire or metal braces. (See Figure 5.)

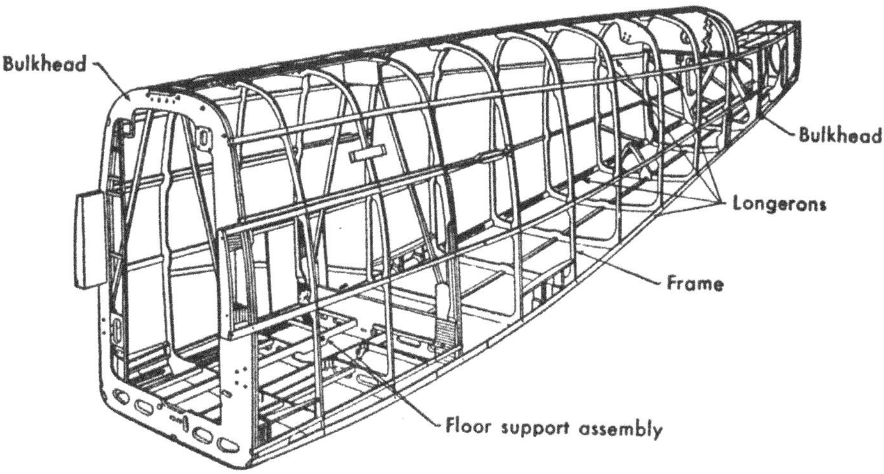

Fig. 7. All-wood fuselage construction. (Courtesy Forest Products Laboratory)

The fuel tanks of most airplanes are located in the wings. The design of the wing depends largely upon the type of airplane and the purpose for which the airplane is to be used.

The tail group, or empennage, is designed almost entirely to give

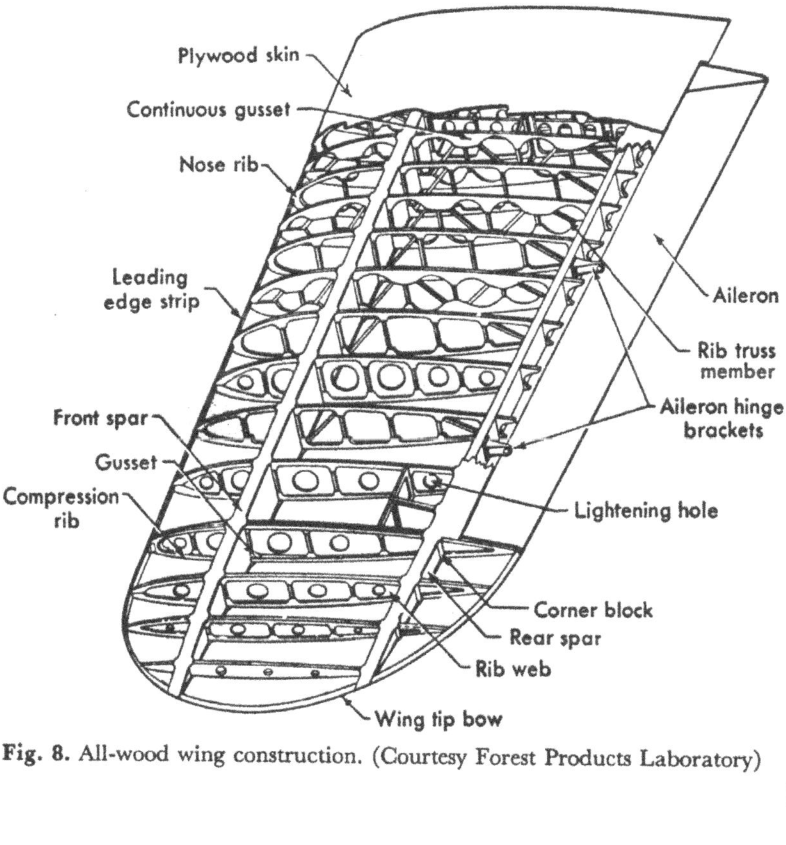

Fig. 8. All-wood wing construction. (Courtesy Forest Products Laboratory)

Fig. 9. A dive bomber showing the cantilever wing. (Courtesy Douglas Aircraft Company, Inc.)

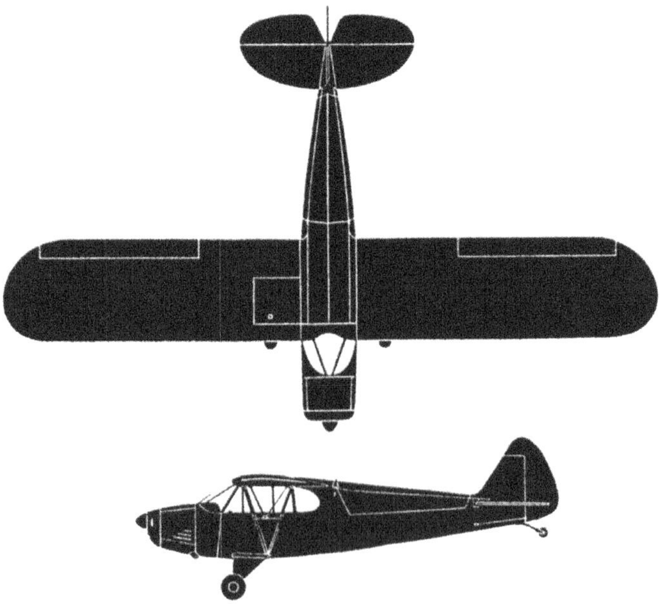

Fig. 10. A plan view and side view of a light airplane. (Courtesy Piper Aircraft Corporation)

control to the airplane in flight and acts somewhat like the feathers on an arrow in maintaining direction of flight. The movable portions, namely, the rudder and elevators, make it possible for the pilot to change the direction of flight.

The landing gears of practically all aircraft are made of metal due to the severe strains placed upon them in taxiing over rough ground

Fig. 11. A four-engine transport airplane. Note the engine nacelles and landing gear. (Courtesy Douglas Aircraft Company, Inc.)

and landing and taking off. Rubber tires, rubber shock cords, and air and oil cylinders are used to reduce shocks to the airplane structure.

When the engine is mounted other than in front of the fuselage, the housing which encloses the engine and its various accessories is called a nacelle. The engine mounts of airplanes are strong, rigid structures, usually of metal. This is necessary because of the strains placed upon the mount by supporting the engine which is turning the propeller at a high rate of speed.

II WOOD
ITS PROPERTIES AND CHARACTERISTICS

Wood is a vegetable growth and forms the main structure of trees. Wood is complex in structure. It is formed of cellulose fibers and other vegetable materials. Cellulose is a material formed in the growing of wood and is very similar to cotton in composition. Although wood has been used since the earliest history of man, there is still much to be learned about its properties and uses. Much research is being done on wood, and new discoveries are constantly being made.

Physical Properties of Wood. Wood varies in color, density, hardness, elasticity, strength, and specific gravity. The density of a substance is

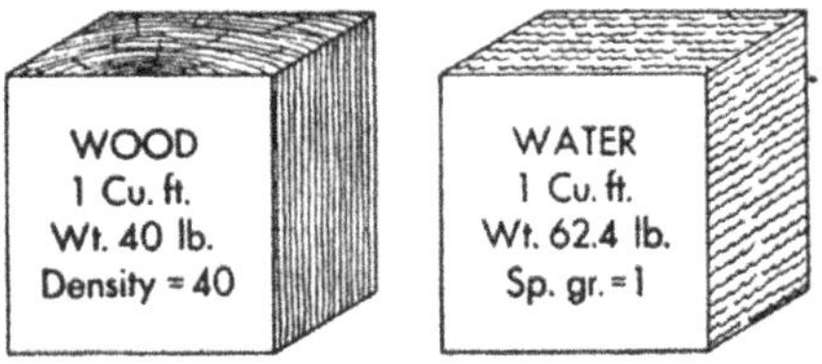

Fig. 12. Density and specific gravity.

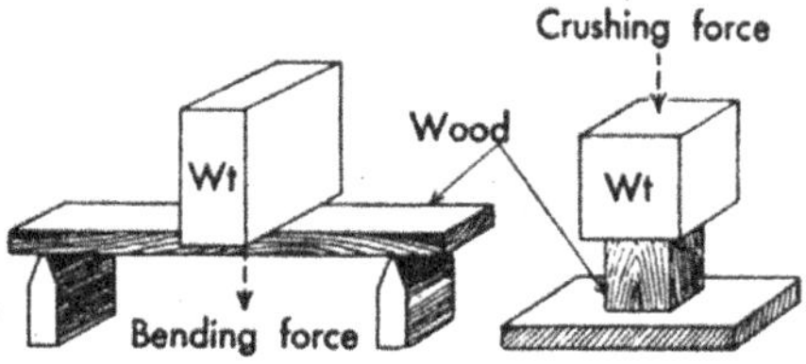

Fig. 13. Bending and crushing forces.

the weight of a unit volume of the substance. For example, the weight of a cubic foot of wood is its density. If a cubic foot of wood weighs 40 lb., the density of that wood is 40. The specific gravity of any substance is the weight of a volume of that substance compared to the weight of the same volume of water. For example, a cubic foot of water weighs 62.4 lb. If a cubic foot of wood weighs 31.2 lb., the specific gravity of that wood is 0.5. The density is the actual weight of a unit volume, and specific gravity is the weight of a unit volume compared to the weight of an equal volume of water. Density and specific gravity should not be confused.

Some woods have their greatest strength in resisting bending forces. Other woods have their greatest strength in resisting crushing forces.

Others show resistance to wear or vibration. Others are very durable, resisting decay under various conditions.

Woods vary in weight from a few pounds per cubic foot to more than 65 lb. per cu. ft. Balsa is the lightest of woods. Dry balsa has a specific

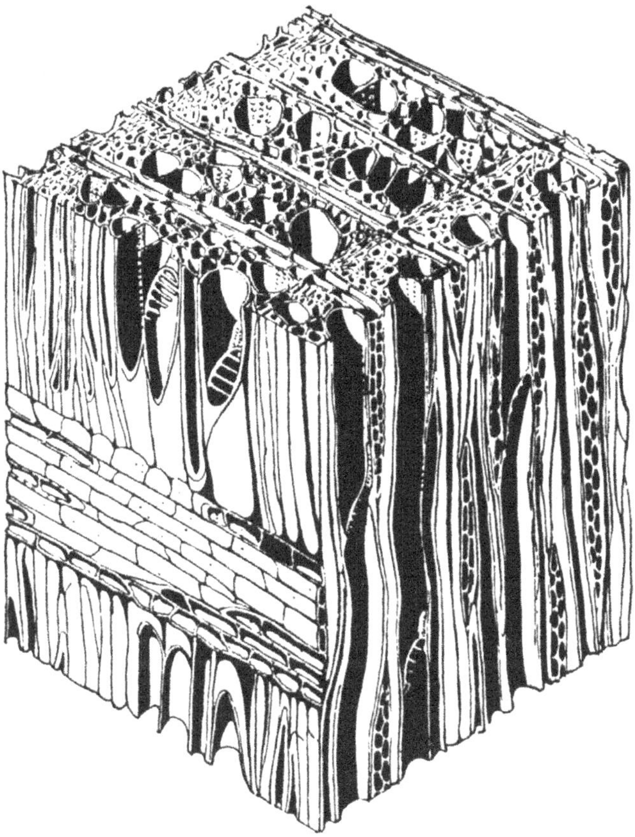

Fig. 14. A magnified cube of wood.

gravity of from 0.12 to 0.20. Its weight per cubic foot varies from 7½ lb. to 12 lb. Some of the tropical woods are very heavy. For example, Ibiraro sometimes weighs as much as 68 lb. per cu. ft. and will sink if placed in water.

Physical Structure of Wood. Wood is made up of a large number of cells. During the growth of the wood these cells are filled with liquids. When wood is dried, these liquids evaporate, leaving the cells light and empty. In addition to the water contained within the cells, from 8 per cent to 12 per cent moisture may be held within the cells or other wood material. A much larger amount of water is held within the cells in the form of sap. In a log freshly cut from a living tree, the water content may weigh from ¼ to 3 times the weight of the oven-dried wood. In the drying of wood, the moisture within the cells is first given up. Wood is either kiln-dried or air-dried. Oven drying is necessary to remove the moisture held in the wood structure. This may amount to 12 per cent of the weight of the oven-dried wood.

The cells are usually long and pointed. These cells are commonly called "wood fibers." The cells vary in length and diameter in various kinds of wood and, to a great extent, determine the properties of the wood. These fibers, which are cellulose, are cemented together by a

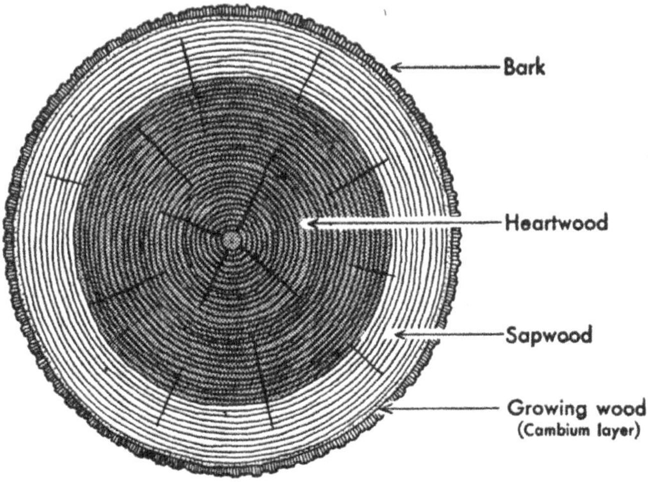

Fig. 15. Cross section of a log showing the various layers.

material called "lignin." Like the cells in the human body, no two are exactly the same size and shape. This can readily be observed by use of a microscope. (See Figure 14.)

Annual Rings. Tree rings, which are the rings seen on the end of a log when a tree is sawed at right angles to the trunk, are called "annual rings." Each ring represents the amount of growth during one year. The thickness of the rings, or, as they are sometimes called, the "grain," shows how much wood the tree added during that year. The rings are not the grain. The grain is the direction in which the wood fibers lie.

Some trees grow very slowly. For example, the mesquite tree, which grows in desert regions, has annual rings so close together that it is impossible to count them with the naked eye. Other trees, particularly those growing in warm, moist climates, have wide annual rings. In these trees the annual rings may be as much as or more than ¼ in. in width. It has been possible by the study of tree rings to trace the wet and dry years back sixty or seventy centuries. During a wet year the tree adds a wide ring, and during dry years the ring is comparatively narrow. Each part of the annual ring has different characteristics. The parts of the annual ring differ in color, strength, hardness, and weight.

The spring growth, which is that part of the ring formed during the moist part of the year when the growth is most rapid, is lighter in color and usually coarser in texture. The part of the ring which is formed during the drier part of the season is darker, denser, usually finer in texture, and stronger. This part of the ring is called the "summer growth." The spring growth is shown by the light-colored part of the ring, and the summer growth by the dark part. Each ring is made up of light and dark wood, but sometimes there is little difference in color between the light and dark parts.

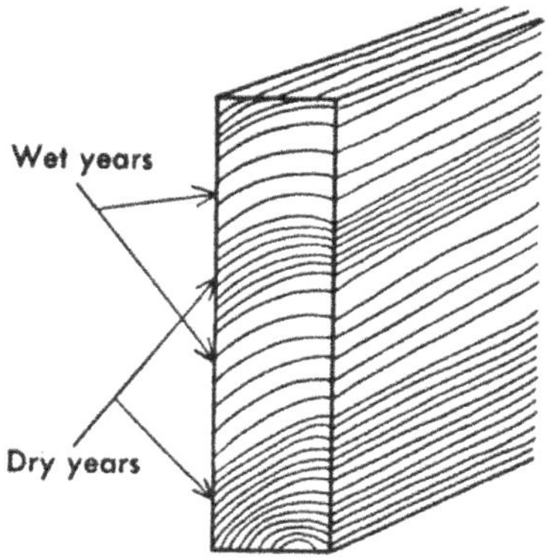

Fig. 16. Variation in annual rings due to wet and dry years.

Heartwood and Sapwood. In different parts of a tree or in a single board the wood often differs in strength, weight, color, and durability. The

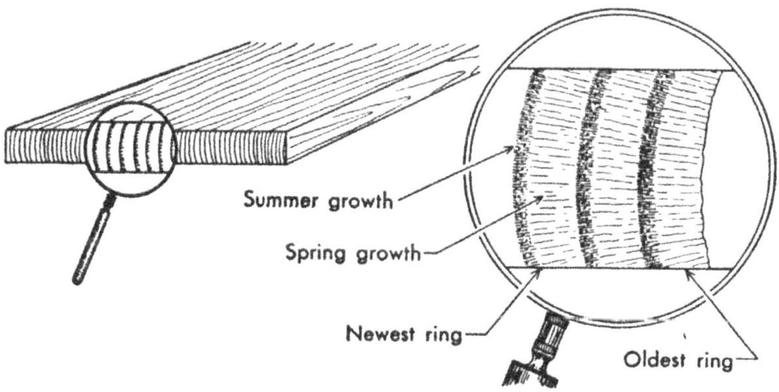

Fig. 17. Annual rings magnified to show spring and summer growth.

older wood is the heartwood or the center of the log. The heartwood is usually darker in color than the wood nearer the bark. There does not seem to be much difference in the strength of the wood whether it comes from the heart or near the bark.

The outer layer of the trunk of a tree contains some sap and is called sapwood. This layer may be several inches thick, depending on the size of the tree. The sapwood is usually lighter in color than the heartwood. (See Figure 15.)

The layer next to the bark is made up of the wood which is being formed. This thin layer is not suitable for use in most structures and is usually removed with the bark when the log is sawed into lumber.

Hardwood and Softwood. Wood is divided roughly into two classes: the hardwoods and the softwoods. One method of classification is that hardwoods are from trees having broad leaves which are shed annually. Softwoods are from those trees having needle-like or scalelike leaves which are not shed annually. The hardwoods usually come from the slower growing trees, for example, oak, maple, and ash, while the softwoods come from the more rapidly growing trees such as redwood, fir, and pine. No classification can be given which definitely separates hardwood and softwood trees. Some trees, particularly those growing in the tropics, are evergreen, and their wood is very hard. Other trees, which are comparatively slow growing and have broad leaves that are shed annually, are soft, for example, basswood and cottonwood.

Physical Characteristics of Wood. The characteristics of wood are divided as follows: durable wood; close grained and open grained wood; strong, medium strong, and weak wood; heavy, medium weight, and light wood. (See Table I.)

TABLE I. CHARACTERISTICS OF SOME COMMON WOODS

KIND	DURABILITY	GRAIN	STRENGTH	HARDNESS	WEIGHT
Birch	Medium	Close	Strong	Hard	Heavy
Hickory	Medium	Open	Strong	Hard	Heavy
Mahogany	Medium	Open	Strong	Hard	Medium
Maple	Poor	Close	Strong	Hard	Heavy
Oak	Good	Open	Strong	Hard	Heavy
Walnut	Good	Open	Strong	Hard	Medium
Cedar	Good	Close	Medium	Medium	Light
Red gum	Poor	Close	Medium	Medium	Medium
Yellow pine	Good	Close	Strong	Medium	Medium
Spruce	Good	Close	Strong	Soft	Light
Basswood	Medium	Close	Weak	Soft	Light
Cypress	Good	Close	Medium	Soft	Light
White pine	Medium	Close	Medium	Soft	Light
Poplar	Good	Close	Medium	Soft	Light
Redwood	Very good	Close	Weak	Soft	Light

Quarter Sawing and Flat Sawing. The way in which a log is sawed largely determines the appearance of the grain and annual rings which show in the finished lumber. Plain-sawed lumber is obtained when the log is sawed by beginning at one side and sawing off boards. The boards which are sawed from the side of a log have a flat grain which does not wear as well as edge grain. As the saw approaches the center of the log the grain becomes more and more perpendicular to the face of the board. The board which is sawed out of the center of the log has the annual rings almost perpendicular to the face of the board. This board is called "edge grain." Edge grain is sometimes called "vertical grain." (See Figure 18.)

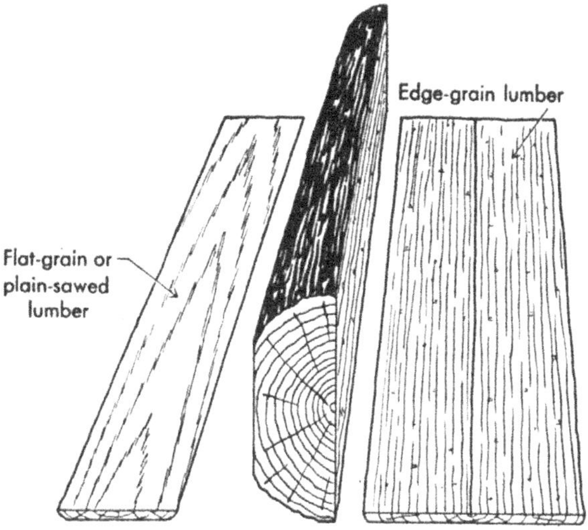

Fig. 18. Flat-grain and edge-grain lumber from the same log.

In quarter sawing the log is sawed in quarters. Each quarter is sawed into boards parallel to the radii of the log. Quarter sawing a log produces all edge-grain lumber.

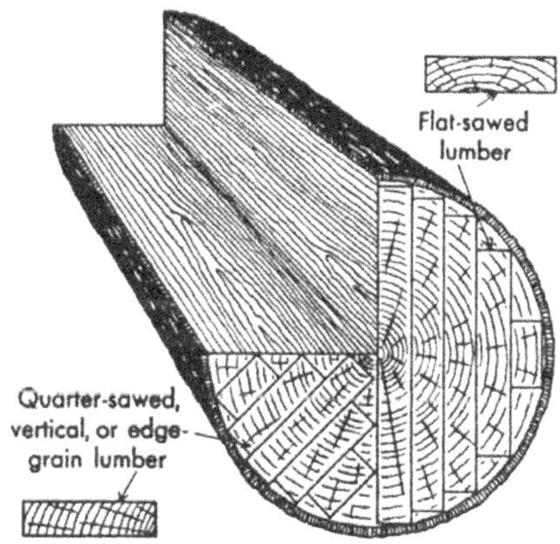

Fig. 19. The method of quarter-sawing lumber.

Lumber Dimensions. For commercial purposes, the dimensions of lumber are given as thickness in inches, width in inches, and length in feet.

Lumber as it comes from the mill is cut in even numbers of inches in width, even numbers of feet in length, and the thickness is measured in full inches. When trees are felled in the forest, the logs are cut in even numbers of feet in length.

Lumber is sold by the board foot. The board foot contains 144 cu. in.

When buying lumber, any thickness less than 1 in. is sold as 1-in. lumber, from 1 to 2 in. as 2-in. lumber, etc. If a board is between 5

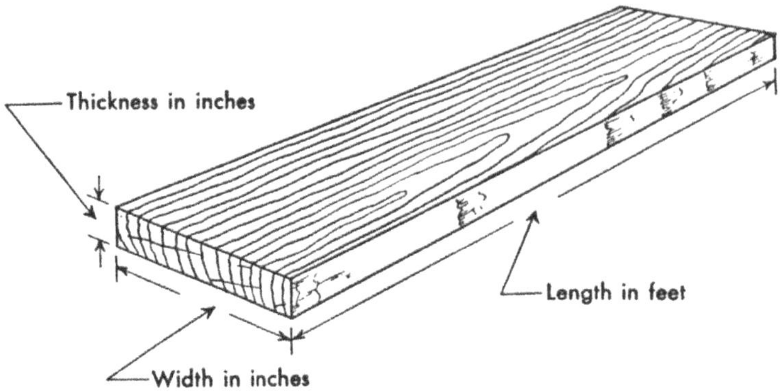

Fig. 20. Conventional lumber measurements.

and 6 in. wide it is sold as 6-in. lumber. A piece of lumber 1½ in. thick, 5½ in. wide, and 10 ft. long would be sold as containing 10 board feet. The number of board feet in a piece of lumber is found by multiplying the thickness in inches by the width in inches, divided by 12, times the feet in length.

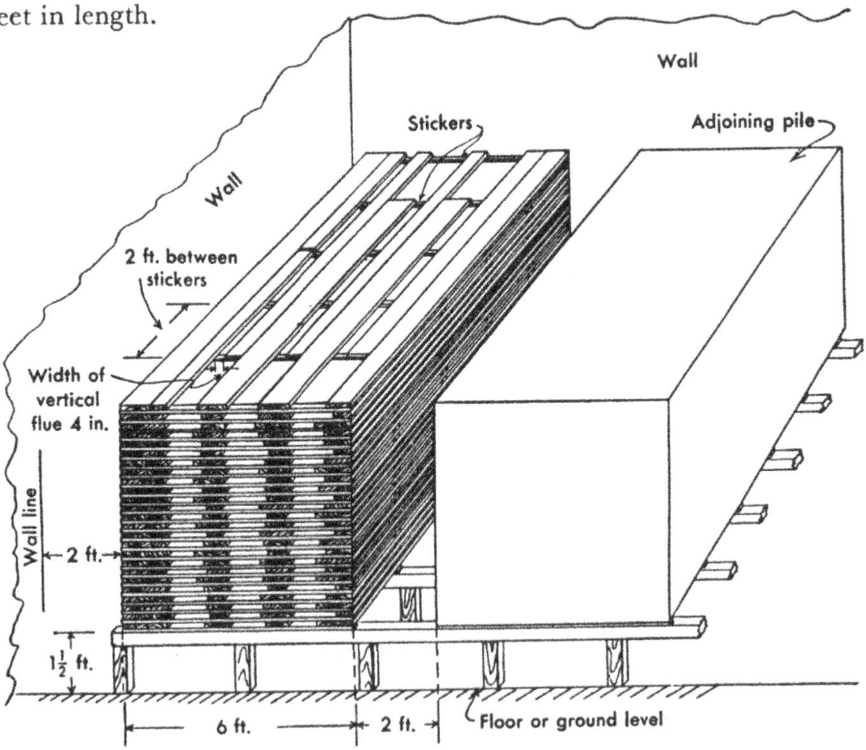

Fig. 21. The proper method of piling lumber. (Courtesy Forest Products Laboratory)

Rough lumber is lumber just as it comes from the saw mill. Finished lumber has had the rough surface removed by planing. Finished lumber, unless otherwise specified, measures less in thickness and width than the size given for that particular piece of lumber. For example, a board sold as 1 × 12 × 10 usually measures approximately ¾ in. thick, 11¾ in. wide, and 10 ft. long, the length being the only dimension which is full size.

Seasoning of Lumber. The treatment which lumber receives after being sawed at the mill has a great deal to do with the characteristics of the wood. Most lumber used in aircraft structures is kiln-dried. Kiln drying is accomplished by placing the freshly sawed lumber in large oven-like rooms called "kilns." The lumber is often treated with live steam or hot water until it is completely penetrated. This treatment accomplishes two purposes. First, it causes an even distribution of moisture throughout the lumber; second, it kills all living matter such as bacteria and insects.

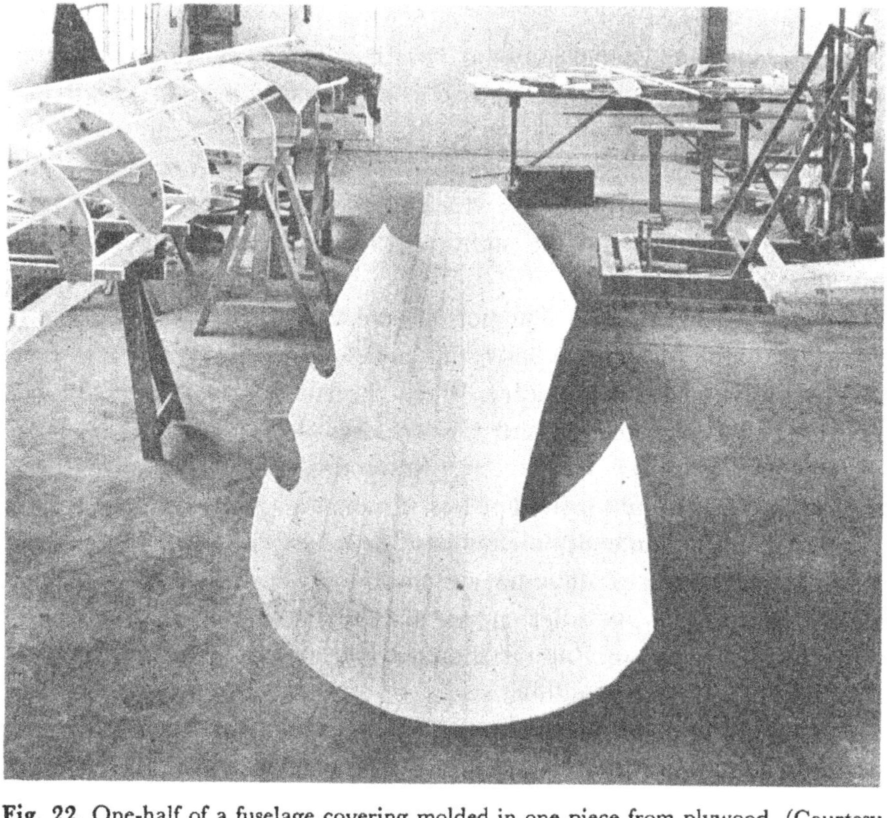

Fig. 22. One-half of a fuselage covering molded in one piece from plywood. (Courtesy Timm Aircraft Corporation)

After being so treated, the wood is dried at a controlled rate by warm air which is passed through the kiln. Controlled drying insures even moisture content. If the rate of drying is too rapid, too great a moisture difference will exist and uneven shrinking will result. Unequal shrinkage may cause stresses in the lumber which manifest themselves in such

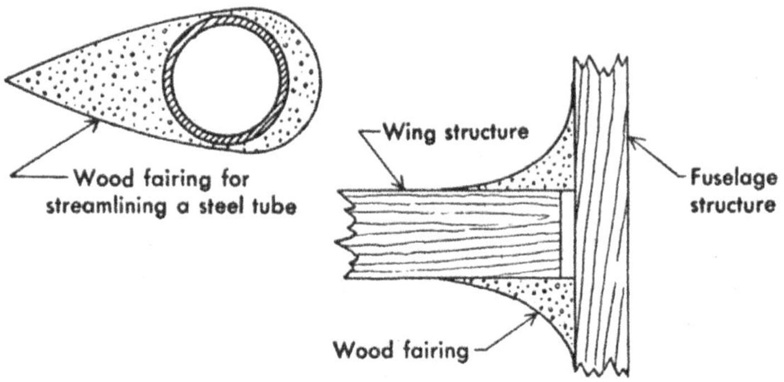

Fig. 23. A light soft wood, such as balsa, is used for fairings and streamlining.

defects as checking, honeycombing, case hardening, or warping. Sometimes the fibers are actually collapsed from too rapid drying. This defect is called "collapse."

Wood that is checked, honeycombed, or collapsed is not suitable for use in aircraft structures. Wood which is case hardened or warped may be used in some parts of the aircraft structure if the defect is not too severe.

Moisture Content of Wood. The moisture content of the wood for aircraft structures should usually be between 8 and 12 per cent. Wood is greatly affected by changes in humidity. When the air is moist, wood absorbs moisture and, when the air is dry, wood loses moisture.

Storing and Handling of Lumber. Aircraft woods should be properly stored to prevent excessive absorption or loss of moisture. Some boards are best stored by standing on end; others should be stored flat with strips placed between the boards to allow for circulation of air. Lumber should not be laid over rafters or other supports which are far enough apart to allow the lumber to sag. Lumber should not be exposed to dust, dropped, or subjected to rough handling.

The handling of lumber is important, as well as its proper storage. Improper handling may cause failures in the board itself and weaken it to such an extent that it will fail after being placed in the airplane structure.

Fig. 24. Preparing plywood for molding. (Courtesy Universal Moulded Products Corporation)

General. It is probable that wood will continue to be used in aircraft construction for many years hence for both private and commercial aircraft. Wood is much cheaper and easier to work and repair than metal. It is used for two general purposes in aircraft construction:

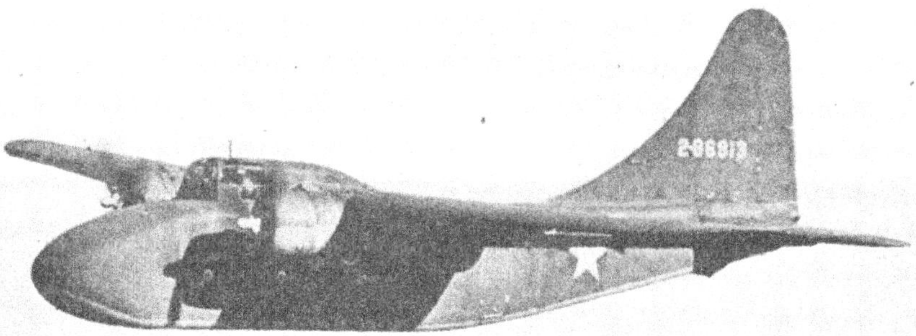

Fig. 25. A modern cargo airplane built almost entirely of wood. The gross weight when loaded is more than 36,000 lb. (Courtesy Curtiss-Wright Corporation)

(1) for structural framework in fuselage and wings; (2) for fairing, streamlining, and covering fuselage and wings.

Due to the shortage brought about by the extensive use of aircraft woods, there is only a limited supply of Grade A material available. Grade A material should not be used where a poorer grade could be used without weakening the structure. Every care should be taken in the selection, inspection, and care of aircraft woods.

One of the large transport planes which has a gross loaded weight of over 36,000 lb. is constructed almost entirely of wood.

III AIRCRAFT WOODS THEIR IDENTIFICATION AND USES

Developments in the use of wood for aircraft construction have been exceedingly rapid in recent years. Although wood is one of the oldest structural materials, modern research is continually developing new, and improving old, wood products.

In general, the desirability of wood for aircraft use depends upon its weight-strength ratio, moisture content, shrinking characteristics, splitting properties, hardness, and toughness. There seems to be at present no real standard by which these properties may be compared for the various woods. These properties vary widely in any kind of wood. For example, some samples of mahogany weigh two or three times as much as other samples of the same species.

The woods used in different parts of any airplane may vary widely in their properties. Spruce, which is suitable for use in all structural parts, is medium light, elastic, strong, tough, and resistant to vibration. However, it is medium soft and not suitable for carrying heavy compression or tension loads.

Oak and maple are hard and strong and are used where heavy compression and tension loads must be supported — for example, in reinforcing blocks and propellers.

Filler blocks may be made of basswood which is not a strong wood but is light, soft, and easily worked.

For fairing, balsa is used because of its extreme lightness. Balsa, however, is a weak wood and is never used for parts of the structure which must carry loads.

The ease with which wood structures may be built and repaired is a factor which has led to the continued use of wood in aircraft construction.

Specific Gravity. A characteristic of wood which is of importance to the aircraft industry is specific gravity. Specific gravity is the number of times a given volume of a substance is as heavy as an equal volume of water. For example, iron is 7.87 times as heavy as water and, therefore, has a specific gravity of 7.87. Aluminum is 2.7 times as heavy as water, and its specific gravity is 2.7. The specific gravity of water is taken as a standard and is 1.0. The specific gravity of most woods is less than 1.0. Specific gravity determines the weight of a given volume of wood. Spruce has an average specific gravity of 0.4. A simple way of determining the specific gravity of a wood is to take a piece of that wood 1 in. square and 10 in. long. This piece is marked off in inch lengths and lowered into a cylinder of water in such a manner that the stick will remain almost upright, as shown in Figure 26. If the stick sinks so that 5.5 of the marks are under the water, the specific gravity is 0.55. If 3 of the marks are under the water the specific gravity is 0.3. If 8.5 of the marks are under the water the specific gravity is 0.85. This method gives a very close approximation to the actual specific gravity.

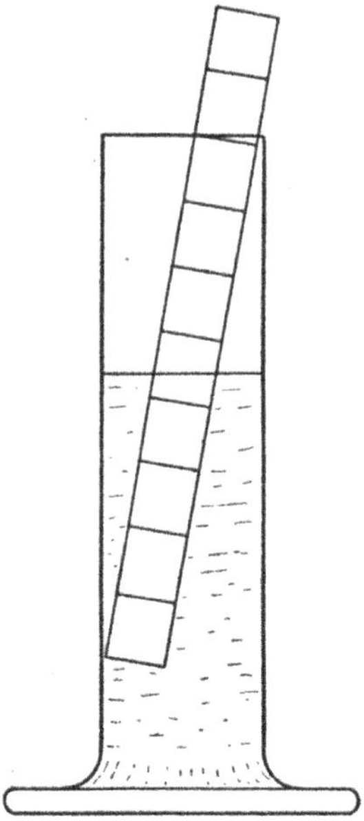

Fig. 26. A method of finding the specific gravity of a wood sample. This figure indicates a specific gravity of approximately 0.45.

Table II roughly classifies aircraft woods as to their strength, moisture content, specific gravity, weight, grain slope, annual rings, strength, and durability.

Hardwoods differ from softwoods in having larger cells or pores scattered among the smaller cells which are the wood fibers. In many hardwoods these pores can be distinctly seen on planed surfaces when the planing is done parallel to the grain. Oak shows these pores plainly, and they are visible to the naked eye as tiny grooves.

Softwoods are composed almost entirely of fine, fibrous cells, lacking the large pores and fibers of the hardwoods. The softwoods do not show, except under magnification, any of the grooves representing the large pores common to hardwoods.

The rays extending from the heartwood toward the bark are much more pronounced in hardwood than in softwood. Most softwood contains much more resin than hardwood. Resin ducts are present in most softwood. These resin ducts may be seen when light is reflected from the planed surface, the resin ducts appearing as shining lines. The direction

TABLE II. MINIMUM REQUIREMENTS FOR AIRCRAFT WOODS

SPECIES OF WOOD	STRENGTH PROPERTIES AS COMPARED TO SPRUCE	PERMISSIBLE RANGE OF MOISTURE CONTENT PER CENT	SPECIFIC GRAVITY (WHEN OVEN DRY)		POUNDS PER CUBIC FOOT AT 15 PER CENT MOISTURE CONTENT	MAXIMUM PERMISSIBLE GRAIN DEVIATION (SLOPE OF GRAIN)	MINIMUM NUMBER OF ANNUAL RINGS PER INCH	REMARKS
			Average	*Minimum*				
Spruce	100%	8–12%	.40	.36	27	1–15	6	Excellent for all uses. Considered as standard for this table.
Douglas fir	Exceed spruce	8–12	.51	.45	34	1–20	8	May be used as substitute for spruce in same sizes or in slightly reduced sizes providing reductions are substantiated. If used as direct substitute for spruce in same sizes, a reduction in the minimum specific gravity to .38 is permissible. Large solid pieces should be avoided. Gluing satisfactory.
Noble fir	Slightly exceed spruce except 8% deficient in shear	8–12	.40	.36	27	1–20		Satisfactory characteristics with respect to workability, warping and splitting. May be used as direct substitute for spruce in same sizes providing shear does not become critical. Hardness somewhat less than spruce. Gluing satisfactory.
Western hemlock	Slightly exceed spruce	8–12	.44	.40	29	1–20	8	Less uniform in texture than spruce. May be used as a direct substitute for spruce. Upland growth superior to lowland growth. Gluing satisfactory.
White pine	Properties between 85% and 96% those of spruce	8–12	.38	.34	26	1–20		Excellent working qualities and uniform in properties but somewhat low in hardness and shock-resisting capacity. Cannot be used as substitute for spruce without increase in sizes to compensate for lesser strength. Gluing satisfactory.
White cedar, Port Orford	Exceed spruce	8–12	.44	.40	30	1–20	8	May be used as substitute for spruce in same sizes or in slightly reduced sizes providing reductions are substantiated. Gluing difficult, but satisfactory joints can be obtained if suitable precautions are taken.
Yellow poplar	Slightly less than spruce except in compression and shear	8–12	.45	.38	28	1–20	8	Excellent working qualities. Should not be used as a direct substitute for spruce without carefully accounting for slightly reduced strength properties. Somewhat low in shock-resisting capacity. Gluing satisfactory.

of the grain can be determined by the resin ducts which extend in the same direction as the wood fibers.

Physical Properties of Wood. Wood is usually identified by means of its physical properties, which are color, weight, hardness, taste, odor, and general structure.

Particular kinds of wood are difficult to describe accurately enough for them to be identified by a written description. The student of woodwork should collect samples of all of the common woods used in aircraft construction and make a study of the characteristics of each wood. These samples need not be large. A piece 6 in. long, 4 in. wide, and 1 in. thick is sufficient. Samples should be taken from both the sapwood and the heartwood. The appearance of the sapwood and heartwood of many species shows distinct differences in their physical properties. These samples should be rough sawed on one side and one edge. The other side and one edge should be planed smooth.

The following properties may be used in the identification of woods:

COLOR. The properties determined by sight are color and grain, or the general appearance of the wood. The color is not always distinctive. Many conditions affect color, such as seasoning, kiln drying, moisture stains, and various types of rot. The color of some woods is distinctive and leads almost at once to their identification. Redwood, mahogany, cedar, and walnut are woods which have distinctive color.

ODOR AND TASTE. Many woods have a characteristic odor or taste. Red cedar and redwood may resemble each other in color, but the well-known cedar smell immediately distinguishes the two. Odors and tastes cannot be accurately described. Each individual must smell and taste a substance in order to recognize it for himself. Taste and odor are usually similar, although some woods may have a sweet, bitter, or sharp taste which is not detected by smell.

WEIGHT. The weight of a wood sample often helps in determining the kind of wood. Many woods vary greatly in weight. Some grades of wood may weigh more than twice as much as lighter grades of the same kind of wood.

HARDNESS. The hardness of a piece of wood is that property which resists cutting or denting, particularly across the grain. Many hardwoods split easily along the grain, while others are tough, and it is almost impossible to get a clean split. This is also true of softwoods. For example, cedar splits readily along the grain, while cottonwood is almost impossible to split cleanly. The hardness of wood may be affected by the moisture content and the amount of seasoning to which it has been subjected. Samples of the same kind of wood may show a decided difference in hardness. Hardness may be defined as the number of pounds pressure necessary to press a steel ball, 0.444 in. in diameter, ½ its diameter into the wood.

It is not necessary for the general worker in wood to be able to identify all woods used in aircraft construction, but he should be familiar with those with which he is apt to work.

The general properties of the more common aircraft woods and their uses are as follows. The weights given are the average for the species when the moisture content is 15 per cent.

Ash. The average specific gravity for white ash is 0.62, minimum permitted 0.56, average hardness 760, and average weight per cubic foot 40 lb. White ash and green ash are very much alike and are not usually distinguished from each other when sold. The sapwood is comparatively light and white. The heartwood is a grayish brown, sometimes tinged with red. Black ash averages considerably lighter than white ash and is not as valuable for aircraft construction. White ash is slightly lighter and softer than oak. It has great strength and stiffness, and bends well. It is particularly suited to construction of bent work, such as wing tips and bulkheads. It is also used in propellers and structural members.

Balsa. The average specific gravity is 0.12 to 0.20, and the weight per cubic foot is from 7½ lb. to 12 lb. This wood has but little strength, is extremely light, and is used for fairings and insulation. It sometimes is impregnated with hot paraffin to make it more workable.

Basswood. The average specific gravity is 0.4, minimum permitted 0.36, average hardness 370, and average weight per cubic foot 22.5 lb. Basswood is usually straight grained, light and solid, with a creamy, light-brown color. The heartwood blends into the sapwood and sometimes has a reddish-brown color. Basswood has a slight characteristic odor even when dry but no distinctive taste. The wood is slightly porous, the pores being invisible without a lens, and has a fine-grained appearance. Basswood is used in aircraft construction in wing rib nose blocks, veneer for plywood, templates, and filler blocks.

Beech. The average specific gravity is 0.67, minimum permitted 0.6, average hardness 1060, and average weight per cubic foot 42 lb. American beech is a hard, heavy wood without characteristic odor or taste. Beech is slightly lighter and stronger than oak, but not as hard. Beech resembles maple. The heartwood of beech has a reddish tinge which blends outward to the sapwood. The sapwood is several inches wide and lighter in color than the heartwood. Beech is used in aircraft construction for plywood veneers and propellers.

Birch. The average specific gravity is 0.59, minimum permitted 0.53, average hardness 1100, and average weight per cubic foot 37 lb. Birch is usually straight grained, without odor or taste, and is approximately the same weight as oak, being slightly stronger but not as hard. The sapwood is wide and light in color, nearly white. The heartwood varies from a deep reddish brown to a light brown tinged with red. The annual rings are quite distinct, and the wood is fairly fine grained, although the pores appear as fine grooves on the dressed surface. White maple is sometimes confused with birch, but the pores are much finer in the maple, which is a distinguishing feature. It is high in wear-resisting qualities. Birch is used in aircraft structures for plywood and propellers.

Cedar (Port Orford). The average specific gravity is 0.44, minimum permitted 0.40, average hardness 520, and average weight per cubic foot 27 lb. Most cedar is straight grained with a characteristic spicy odor and taste. Most cedars are pale brown in color and are quite soft and often slightly spongy in structure. Red cedar is straight grained, light weight, has a reddish-brown color, with a distinct cedar smell and bitter taste. Annual rings are distinct, and the summer wood is well defined and rather thin. Port Orford cedar may be substituted for spruce. Other cedars are not suitable for aircraft construction.

Cherry (Black). The average specific gravity is 0.54, minimum permitted 0.48, average hardness 900, and average weight per cubic foot 33 lb. Black cherry is moderately heavy, fairly straight grained, and without characteristic odor or taste. The sapwood is narrow and blends into the heartwood, which has a reddish-brown, lustrous color. Black cherry is easily distinguished by its color. It resembles mahogany but does not have the large pores characteristic of mahogany. It is used in aircraft construction for propellers and plywood facing.

Cottonwood. The average specific gravity is 0.43, minimum permitted 0.39, average hardness 410, and average weight per cubic foot 27 lb. Cottonwood is not usually straight grained. It is moderately light, soft, and subject to warping. Cottonwood has a slight characteristic odor and is without taste. The heartwood is light greenish brown and not clearly separated from the sapwood. The sapwood is a grayish white in color. Its use in aircraft construction is in veneer for plywood cores.

Elm. The average specific gravity is 0.65, minimum permitted 0.6, average hardness 1230, and average weight per cubic foot 41 lb. Elms are moderately heavy, and the wood works easily. American elm varies from light and brash to heavy and tough. Rock elm and winged elm are heavier, harder, and have better qualities than the American elm. The heartwood is brown with a reddish tinge. Interlocked grain, which results in a tendency to warp or shrink, is common in elm wood. The heart of the log produces lumber which is rather light in color but has a slightly brownish tinge. Elm wood is odorless and tasteless. This wood is not commonly used in aircraft construction but, when used, is substituted for ash in stringers, longerons, and bent work.

Firs. The average specific gravity is 0.41, minimum permitted 0.37, average hardness 450, and average weight per cubic foot 26 lb. The true firs, as distinguished from Douglas fir, are much lighter in color, some of them being nearly white. The spring wood is much lighter than the summer wood which is apt to have a lavender tinge or to be slightly brownish. The true firs closely resemble western hemlock. They are not as hard, heavy, or as strong as Douglas fir. They are moderately light to light in weight and have no characteristic odor or taste. They may be substituted for spruce in aircraft construction but are somewhat heavier.

Fir (Douglas). The average specific gravity is 0.51, minimum permitted 0.45, average hardness 620, and average weight per cubic foot 32 lb. Douglas fir is more durable, stronger, heavier, and more resinous than the other firs. The reddish hue of the heartwood is quite distinctive. Douglas fir has a typical fir odor when being worked. The sapwood is usually several inches wide. The annual rings are very distinct, with many resin ducts present. Douglas fir resembles western larch, but is generally closer grained. It is used in aircraft construction as a substitute for spruce, being stronger but considerably heavier.

Hemlock. The average specific gravity is 0.44, minimum permitted 0.40, average hardness 650, and average weight per cubic foot 28 lb. Hemlock is usually straight grained, moderately light, and pale brown in color. The sapwood blends into the slightly browner color of the heartwood which is apt to have a reddish tinge. The annual rings are clearly defined. Hemlock resembles white fir. It is used in aircraft construction as cores for plywood and for filler blocks.

Hickory. The average specific gravity is 0.8, minimum permitted 0.7, average hardness 1250, and average weight per cubic foot 50 lb. True hickory is tough, hard, and heavy. It is heavier and stronger than oak and is about the same in hardness. It is without characteristic odor or taste. The sapwood is several inches in thickness and has a pale yellow tinge. The heartwood is reddish brown. The annual rings are clearly defined. Hickory is not easily confused with other woods. The distinct annual rings and great hardness is characteristic of this wood. It is useful in aircraft construction for compression blocks and stringers.

Mahogany. The average specific gravity is 0.51, minimum permitted 0.46, average hardness 790, and average weight per cubic foot 32 lb. Mahogany is considerably lighter and softer and slightly below oak in strength. The reddish-brown appearance of mahogany, which turns darker on exposure to light, is one of the characteristic properties of this wood. The pores in mahogany are visible without magnification. Mahogany may be confused with Spanish cedar but does not have the same characteristic taste or odor. Different samples of mahogany vary considerably in weight. Mahogany is used largely in aircraft construction for propellers and plywood.

Maple. The average specific gravity is 0.68, minimum permitted 0.6, average hardness 1270, and average weight per cubic foot 42 lb. Sugar maple and black maple are difficult to cut across the grain, are hard and heavy, and are classified as hard maples. Red maple and silver maple are classed as soft maple, and neither is hard or heavy. Maple has a reddish-brown heartwood without odor or taste. Annual rings are thin and reddish in color, usually easily seen on planed surfaces. Maple is slightly lighter and harder than oak. Maple is used in aircraft construction for propellers, plywood veneer, and engine mounts.

Oak (Red). The average specific gravity is 0.67, minimum permitted 0.62, average hardness 1240, and average weight per cubic foot 42 lb. The red-oak group is approximately the same in weight and hardness as the white-oak group. The sapwood varies from one to three inches in width. A reddish tinge is usually present in the heartwood. Some pieces of red oak may resemble white oak rather closely in color. The dry wood is without characteristic odor or taste. The green wood has a sour odor. The annual rings of red oak are usually wider and more distinct than those of white oak. Red oak is used in aircraft structures for propellers, engine mounts, and compression blocks.

Oak (White). The average specific gravity is 0.72, minimum permitted 0.62, average hardness 1240, and average weight per cubic foot 45 lb. The wood of the white-oak group is hard and heavy. The sapwood is from one to two inches wide; the heartwood is grayish brown and does not ordinarily have any tinge of red. Oak when dry is without a distinct odor or taste, although persons familiar with this wood can recognize the odor as the wood is being sawed or planed. Green oak has a sour odor. A characteristic feature of all oak wood is the presence of little grooves on the planed surface. The presence of these grooves is due to the large pores in the wood. Broad rays on the end surface which appear on the planed surface as silvery patches are characteristic of this wood. Plane-sawed white oak may be confused with chestnut, but the chestnut is lighter in weight and has finer rays. The oaks cannot be substituted for spruce. White oak is used in aircraft structures for propellers, engine mounts, and compression blocks.

Poplar (Yellow). The average specific gravity is 0.43, minimum permitted 0.38, average hardness 420, and average weight per cubic foot 27 lb. Yellow poplar is usually straight grained, without characteristic odor or taste, and is moderately light. The light-to-moderately-dark, yellowish-brown heartwood has a greenish tinge and sometimes is discolored a purplish-brown color. The annual rings are not distinct, being merely light-colored lines. The sapwood is a yellowish white. Yellow poplar is used in aircraft construction in plywood cores and nose blocks for ribs.

Pine (White). The average specific gravity is 0.37, minimum permitted 0.34, average hardness 380, and average weight per cubic foot 23 lb. White pine has a slight resinous taste and a slight but distinct resinous odor. It is light to moderately light in weight, and straight grained. The heartwood is creamy brown to a light reddish brown, and the knots are usually reddish in color. The summer wood is not distinct from the spring wood. Pitch or resin sometimes appears on the surface of the wood when the wood is heated. White pine is used in aircraft construction for webs, cap strips, corner blocks, fairing strips, patterns, and forms.

Pine (Yellow). The average specific gravity is 0.36, minimum permitted 0.32, average hardness 380, and average weight per cubic foot 26 lb. Yellow pine is more resinous than white pine and has a much wider and harder layer of summer wood. The yellow pine is harder and heavier than white pine. The other characteristics are similar to white pine except that the color is tinged with a yellowish brown, with a reddish-brown to orange-brown heartwood. It may be substituted in aircraft structures for white pine.

Spruce. The average specific gravity is 0.40, minimum permitted 0.36, average hardness 440, and average weight per cubic foot 27 lb. The spruces are usually straight grained. They are moderately light and are used extensively in aircraft structures. In red and white spruce the heartwood is as light as the sapwood. Spruces are pale golden white in color. There is often a light reddish tinge to the heartwood of Sitka spruce which makes it slightly darker than the sapwood. Annual rings are clearly defined. Spruce is apt to be confused with pine, particularly white pine. Spruce is distinguished from white pine by its more prominent annual rings. Split or smoothly dressed surfaces of Sitka spruce usually have slight indentations or a "pocked" appearance never found in Douglas fir or white pine. Some pines do, however, have a pocked appearance, particularly the western and yellow pine. Spruce is used in aircraft construction in many structural parts such as spars, struts, longerons, float and hull structural members, construction of ribs, webs, landing gear, cap strips, stiffeners, flooring, planking, and plywoods.

Walnut (Black). The average specific gravity is 0.56, minimum permitted 0.52, average hardness 990, and average weight per cubic foot 35 lb. Black walnut is heavy and hard and usually straight grained. The characteristic, rich, chocolate-brown color of the heartwood is dinstinctive. The sapwood is lighter in color but tinged with the same chocolate-brown color. The annual rings are fairly distinct and are characterized by medium-large pores in the spring wood. The color and distinct pores distinguish black walnut from other woods. It is resistant to shock and wear and is easily worked. It does not dry well but, if properly dried, holds its shape and is suitable for both bent and flat work. Its use in aircraft structures is for propellers and outside veneers of plywood.

IV DEFECTS IN AIRCRAFT WOODS

Every woodworker, whether he is building a new structure or making repairs, must be an inspector of wood. It is part of his job to determine whether or not the material at hand is suitable for its particular use in the aircraft. It is just as important for the repair man to select the proper grade of material as it is for the original builder. Many defects are discovered only when the wood is being worked.

When a log is sawed, it almost always produces several grades of lumber. A few boards may be nearly perfect. These pieces are graded as No. 1. Some pieces are graded as cull lumber. Cull is the poorest grade of lumber and has practically no use in aircraft structures.

There may be such imperfections as grain irregularities, knots, checks, decay, and compression failure.

Grain. There are several grain defects, or variations in grain, which may cause an apparently sound piece of wood to be rejected for aircraft use.

The direction of the grain is very important. Although the grain itself may be straight, it may cross the piece of wood at such an angle as to cause the material to be classified as defective. Most wood tends to split along the grain. The grain itself is the direction in which the wood fibers lie. Any variation from straightness of grain weakens the material. Grain should always be taken into consideration before using the wood in an aircraft part. Grain is hardly ever perfect, so certain variations are allowed.

Slope of grain is given as the number of inches it varies from a center line in a given distance. For example, if the grain deviates 1 in. in 10 in. the grain slope is 1 in 10. Steeper slopes would be less than 1 in 10 — for example, 1 in 5. A slope of 1 in 20 is not as steep as a slope of 1 in 10. The steeper the grain slope, the weaker the piece of wood in resisting bending forces. Perfect grain would be a grain which does not vary from the center line of the piece of lumber. This grain would run straight through from one end of the piece to the other.

Grain may not only vary from a straight line but may take a spiral or twisting course. This is called "spiral grain." Sometimes variation of grain is caused by sawing tapered logs. When logs of this kind are sawed, the grain may run diagonally through the board. Diagonal grain

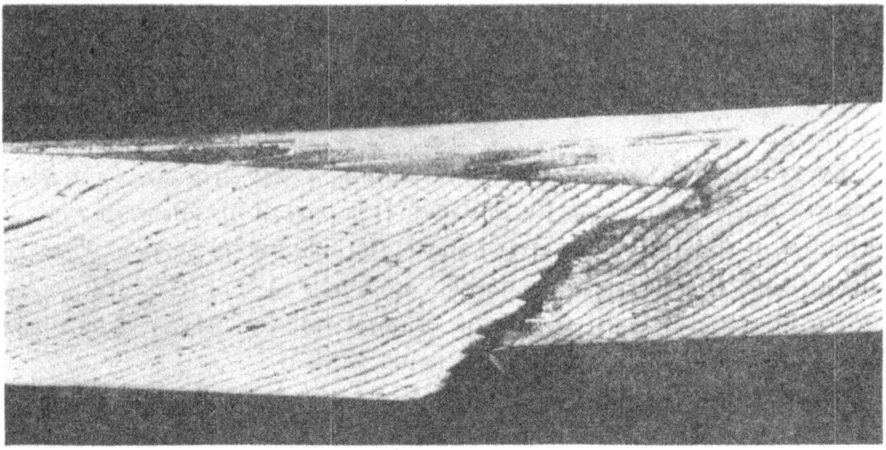

BOTTOM VIEW

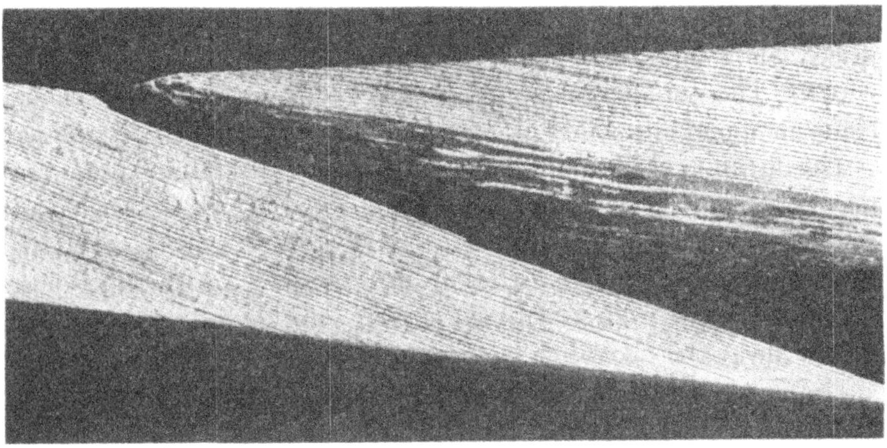

SIDE VIEW

Fig. 27. A typical cross-grain break. (Courtesy Forest Products Laboratory)

can usually be determined by the appearance of annual rings along the smooth surface of the board as shown in Figure 30.

The larger the number of annual rings appearing along the side of the board, the greater the slope of the grain. Some grain is not only sloping or spiral but may actually be curly or twisting. This kind of lumber has a wide variation in grain direction. The grain may run in a number of different directions in a few inches of length. Grain direction can usually be determined by dropping a little free-flowing ink on the surface of the wood. The ink will flow along the grain direction. (See Figure 35.) In pieces which will be subjected to great stresses, a slope greater than 1 in 20 should not be used.

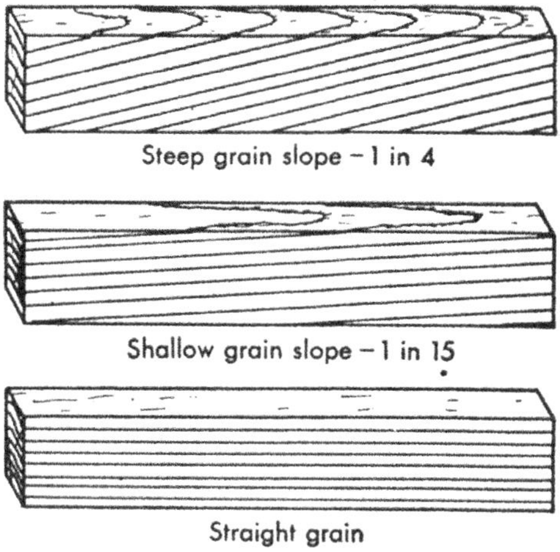

Fig. 28. Grain slope.

Fig. 29. Spiral grain shown by the appearances of the annual rings on the surface of the board. (Courtesy Forest Products Laboratory)

Knots. A knot is a part of a branch which has become embedded in the wood during its growth. The tree grows in size by adding a layer of wood each year. This added layer covers up a certain portion of the branch. A branch which started when the tree was small becomes embedded in the wood, forming a knot. The chief objection to knots

Fig. 30. Diagonal grain shown by the appearance of the annual rings on the surface of the board. (Courtesy Forest Products Laboratory)

is the local effect they have on grain. Knots cause the grain to vary from a straight line, as shown in Figure 36. The grain does not go through the knot but around it. If knots are tight in the wood they are sometimes allowed in certain parts of aircraft structures. Table III shows allowable defects and defects which are not allowable, to assist in determining whether the piece of wood containing the knot may be safely used.

Compression Wood. When a tree, during the time of its growth, leans from the vertical, compression wood is formed on the lower side of the tree trunk. The side toward which the tree leans contains the compression wood. It seems that the tree tries to reinforce this part of its structure to carry the extra strain. The annual rings are much thicker in this

Fig. 31. Double spiral growth of the grain of a tree. (Courtesy Forest Products Laboratory)

Fig. 32. Spiral grain shown in lumber and in the trunk of a tree. (Courtesy Forest Products Laboratory)

part of the log. This wood is heavy, weak, and subject to extreme shrinking. Figure 38 shows compression wood as it appears in the end of a log and in the board after sawing. Compression wood is not allowed in aircraft structural parts which must carry heavy loads. Compression wood may be recognized by the wide annual rings. This wood may have a lifeless appearance.

Fig. 33. Curly grain as shown on the surface of lumber. (Courtesy Forest Products Laboratory)

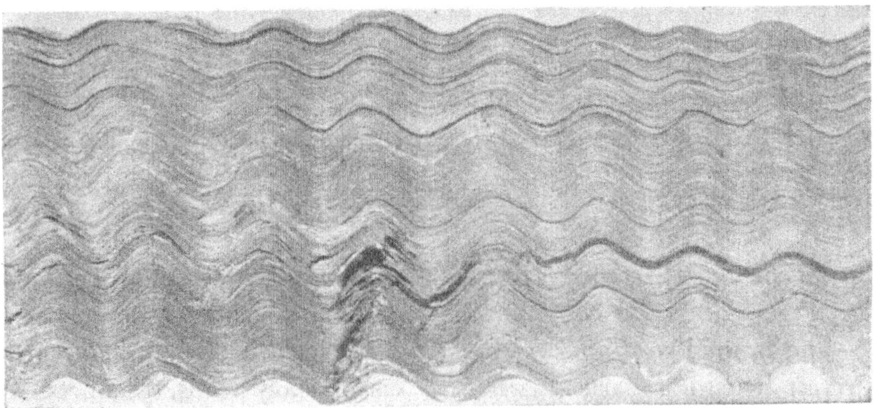

Fig. 34. Curly grain as shown on a split surface. (Courtesy Forest Products Laboratory)

Fig. 35. The ink flow indicates the grain direction.

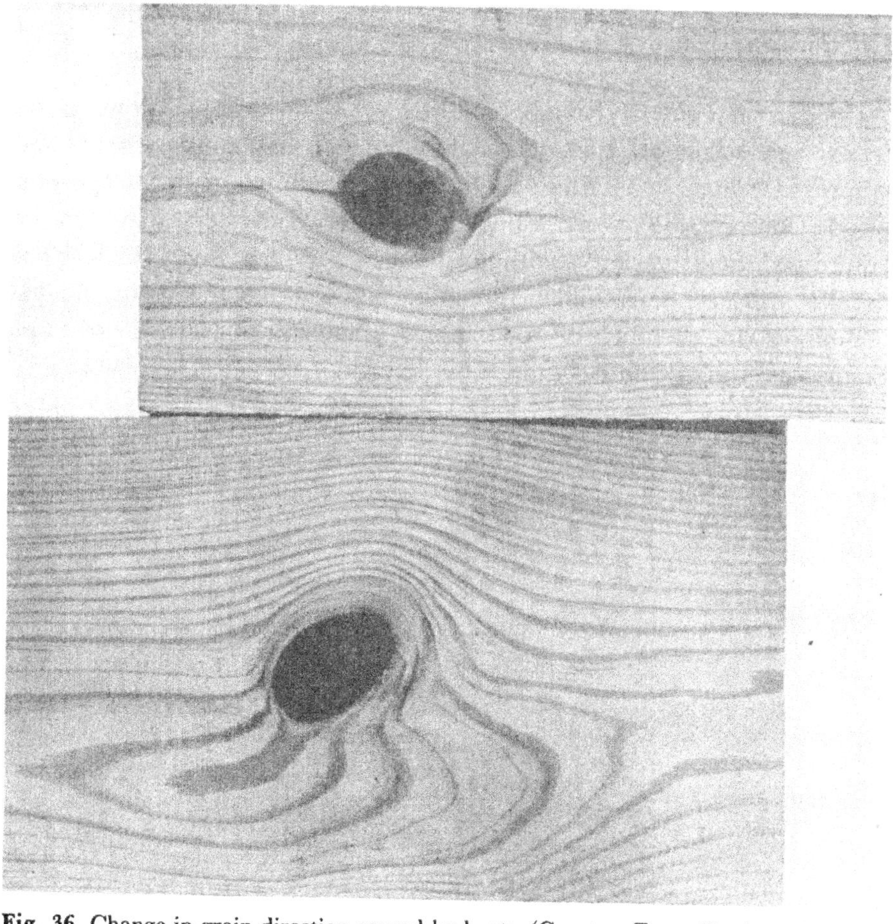

Fig. 36. Change in grain direction caused by knots. (Courtesy Forest Products Laboratory)

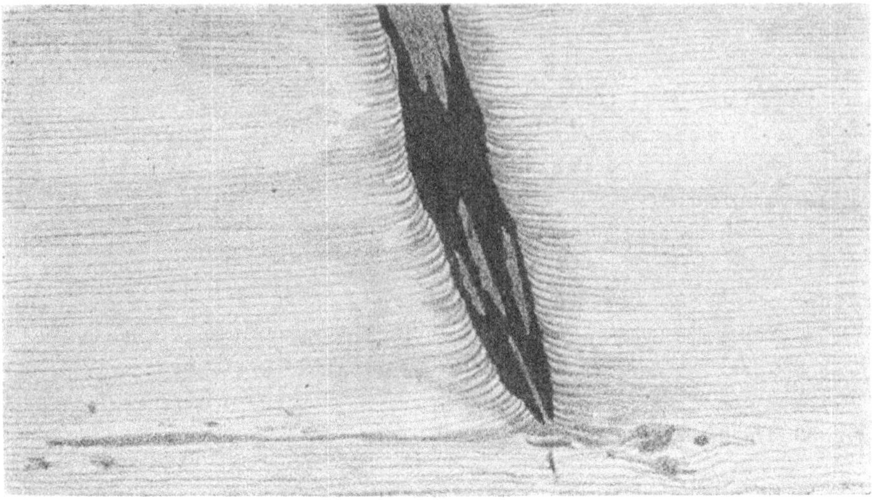

Fig. 37a. Spike knot.

Compression Failures. Rough handling of logs, such as allowing the tree to fall across a large rock or another log, may cause compression failures. Compression failure must not be confused with compression wood. Compression failure is buckling of the wood fibers. This type of failure is difficult to detect without the use of a microscope. This defect usually appears as irregular, threadlike lines across the grain. Under the microscope the fibers of wood appear to have been crushed endways. Figures 39 and 40 show compression failures.

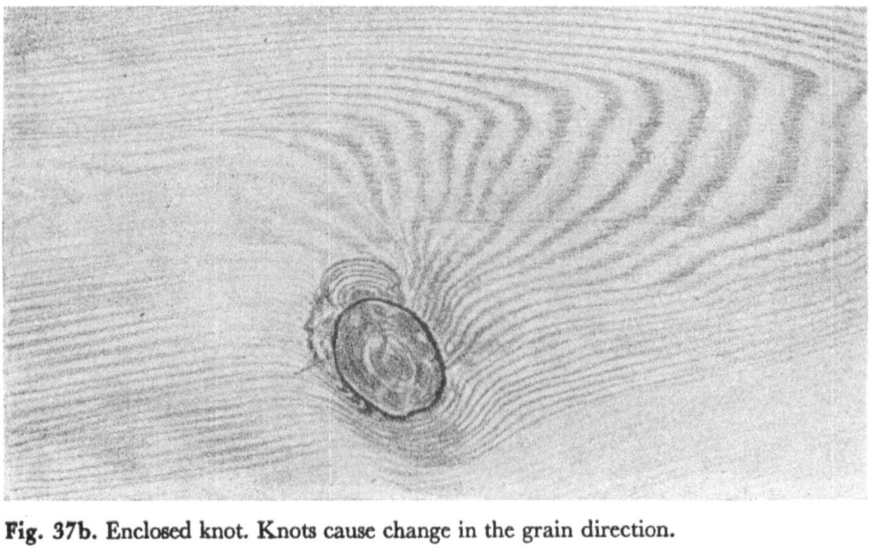

Fig. 37b. Enclosed knot. Knots cause change in the grain direction.

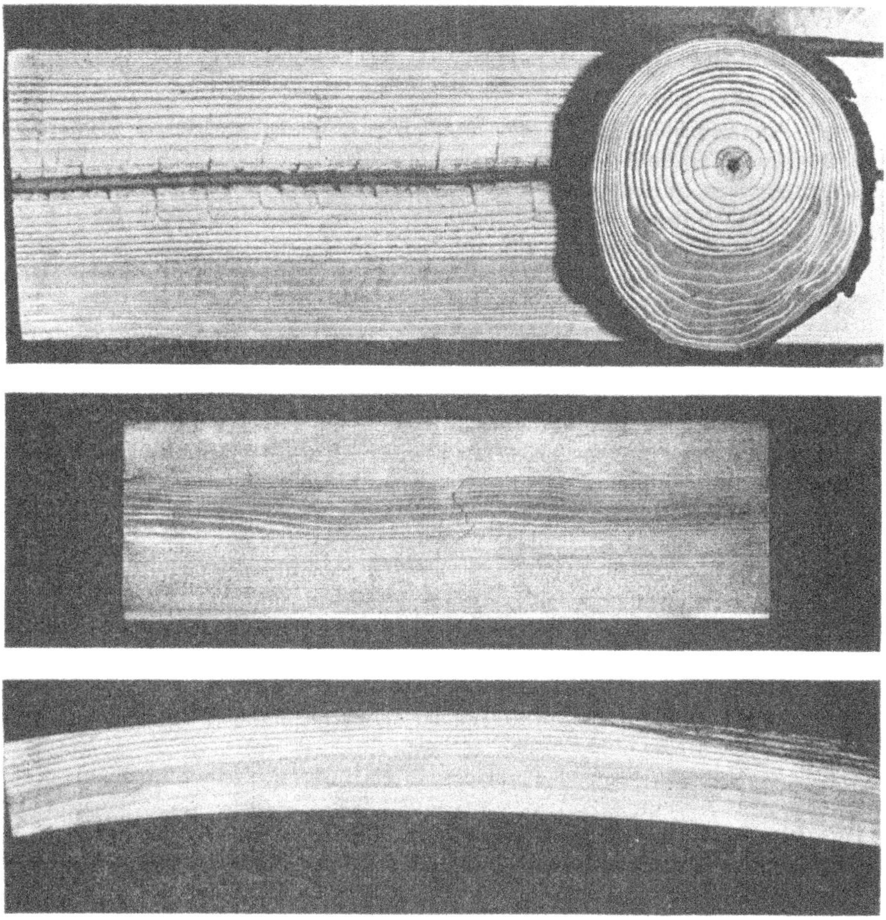

Fig. 38. Compression wood in a log and in sawed lumber. (Courtesy Forest Products Laboratory)

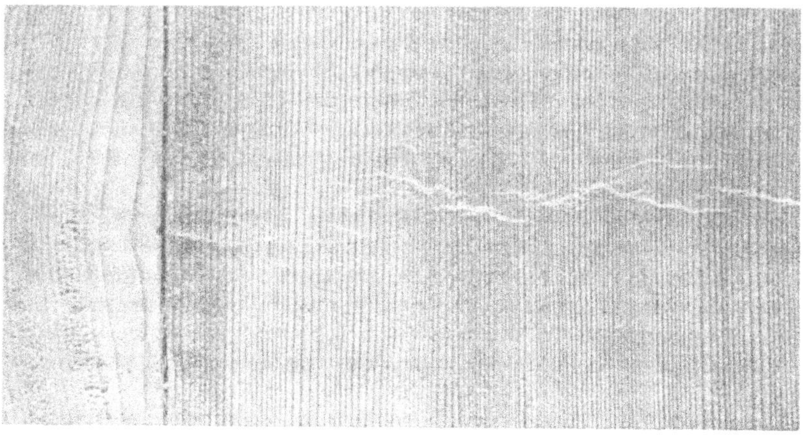

Fig. 39. Compression failure. (Courtesy Forest Products Laboratory)

TABLE III. DEFECTS PERMITTED AND NOT PERMITTED

1. DEFECTS PERMITTED

(a) *Cross Grain.* Spiral grain, diagonal grain, or a combination of the two is acceptable providing the grain does not diverge from the longitudinal axis of the material more than specified. A check of all four faces of the board is necessary to determine the amount of divergence. The direction of free-flowing ink will frequently assist in determining grain direction.

(b) *Wavy, Curly, and Interlocked Grain.* Acceptable if local irregularities do not exceed limitations specified for spiral and diagonal grain.

(c) *Hard Knots.* Sound hard knots up to ⅜ in. in maximum diameter acceptable providing (a) they are not in projecting portions of I-beams, along the edges of rectangular or beveled unrouted beams, or along the edges of flanges of box beams (except in lowly stressed portions); (b) they do not cause grain divergence at the edges of the board or in the flanges of a beam more than specified; and (c) they are in the center third of the beam and are not closer than 20 in. to another knot or other defect (pertains to ⅜-in. knots — smaller knots may be proportionately closer). Knots greater than ¼ in. should be used with caution.

(d) *Pin-Knot Clusters.* Small clusters acceptable providing they produce only a small effect on grain direction.

(e) *Pitch Pockets.* Acceptable in center portion of a beam providing they are at least 14 in. apart when they lie in the same growth ring and do not exceed 1½ in. length × ⅛ in. width × ⅛ in. depth and providing they are not along the projecting portions of I-beams, along the edges of rectangular or beveled unrouted beams, or along the edges of the flanges of box beams.

(f) *Mineral Streaks.* Acceptable providing careful inspection fails to reveal any decay.

2. DEFECTS NOT PERMITTED

(a) *Cross Grain.* Not acceptable unless within limitations noted in 1a.

(b) *Wavy, Curly, and Interlocked Grain.* Not acceptable unless within limitations noted in 1b.

(c) *Hard Knots.* Not acceptable unless within limitations noted in 1c.

(d) *Pin-Knot Clusters.* Not acceptable if they produce large effect on grain direction.

(e) *Spike Knots.* These are knots running completely through the depth of a beam perpendicular to the annual rings and appear most frequently in quarter-sawed lumber. Wood containing this defect should be rejected.

(f) *Pitch Pockets.* Not acceptable unless within limitations noted in 1e.

(g) *Mineral Streaks.* Not acceptable if accompanied by decay. (See 1f.)

(h) *Checks, Shakes, and Splits.* Checks are longitudinal cracks extending, in general, across the annual rings. Shakes are longitudinal cracks usually between two annual rings. Splits are longitudinal cracks induced by artificially induced stress. Wood containing these defects should be rejected.

(i) *Compression Wood.* This defect is very detrimental to strength and is difficult to recognize readily. It is characterized by high specific gravity, has the appearance of an excessive growth of summer wood, and in most species shows but little contrast in color between spring wood and summer wood. In doubtful cases, the material should be rejected or samples should be subjected to a toughness machine test to establish the quality of the wood. All material containing compression wood should be rejected.

(j) *Compression Failures.* This defect is caused from the wood being overstressed in compression due to natural forces during the growth of the tree, felling trees on rough

TABLE III. DEFECTS PERMITTED AND NOT PERMITTED *(continued)*

or irregular ground, or rough handling of logs or lumber. Compression failures are characterized by a buckling of the fibers that appears as streaks on the surface of the piece substantially at right angles to the grain, and they vary from pronounced failures to very fine hairlines that require close inspection to detect. Wood containing obvious failures should be rejected. In doubtful cases, the wood should be rejected or further inspections in the form of microscopic examination or toughness tests made, the latter means being the more reliable.

(k) *Decay.* All stains and discolorations should be examined carefully to determine whether or not they are harmless stains or preliminary or advanced decay. All pieces should be free from rot, dote, red heart, purple heart, and all other forms of decay.

Checks, Shakes, and Splits. Checks, shakes, and splits sometimes resemble each other rather closely. They all seriously lower the strength of the wood, and the defect is pronounced.

A check is a crack crossing the annual rings. Checks are usually caused by faulty drying or seasoning. Checks follow the same direction as the radii of the log from which the lumber was sawed.

A shake is a crack appearing between the annual rings. A shake is formed by the annual rings breaking apart. This may be found in either green or dry wood. A shake always follows the annual-ring direction.

A split is a lengthwise separation of the wood and is caused by the tearing apart of the wood cells. A split is commonly called a crack. A crack does not necessarily follow the annual rings or the radii of the log but usually follows the grain direction. Splits may be caused by excessive stresses in the wood or by shrinking due to changes in moisture content.

Pitch Pockets. Pitch pockets are openings in the wood, usually occurring between the annual rings, which may resemble shakes. These openings are filled with resin or pitch which may be either solid or semiliquid. Pitch pockets sometimes contain particles of the bark and may vary greatly in size. Pitch pockets always weaken the lumber. Small pitch pockets, although not usually objectionable, may weaken a piece of lumber enough to cause its rejection. Changes in temperature may cause the pitch to flow out on the surface of the part, thereby destroying the protective finish.

Bark Pockets. Sometimes an injury to a portion of a tree may kill the living layer next to the bark. The rest of the tree continues to grow, and the wood forms over the injured portion, enclosing pieces of bark. This condition is called a bark pocket. Bark pockets always weaken the wood and are objectionable.

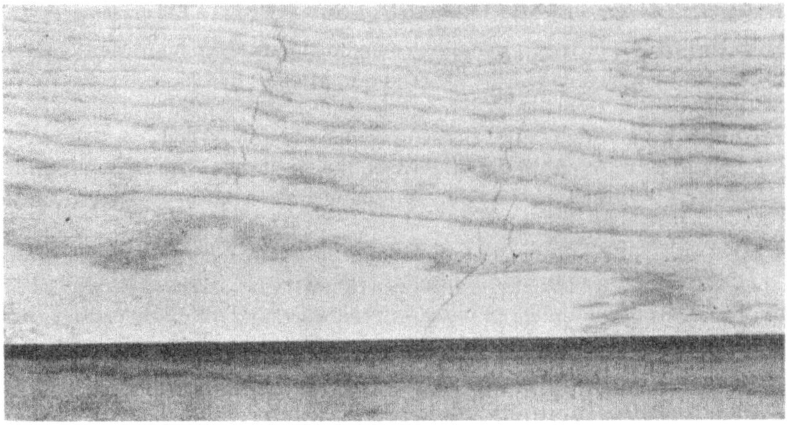

Fig. 40. Compression failure. (Courtesy Forests Products Laboratory)

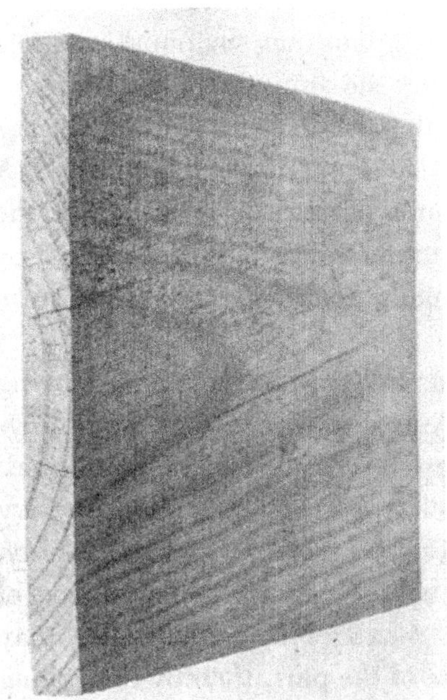

Fig. 41. Checks in the end of a plank. (Courtesy Forest Products Laboratory)

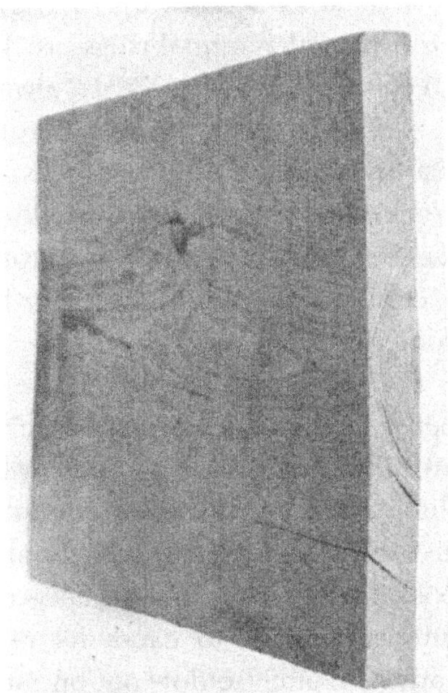

Fig. 42. Shakes in the end of a plank. (Courtesy Forest Products Laboratory)

Pitch Streaks. Pitch streaks should not be confused with pitch pockets. Pitch streaks appear as thick annual rings. These streaks are caused by pitch being absorbed into the wood fibers. Pitch pockets contain pure pitch, while pitch streaks are formed of wood soaked in pitch. These

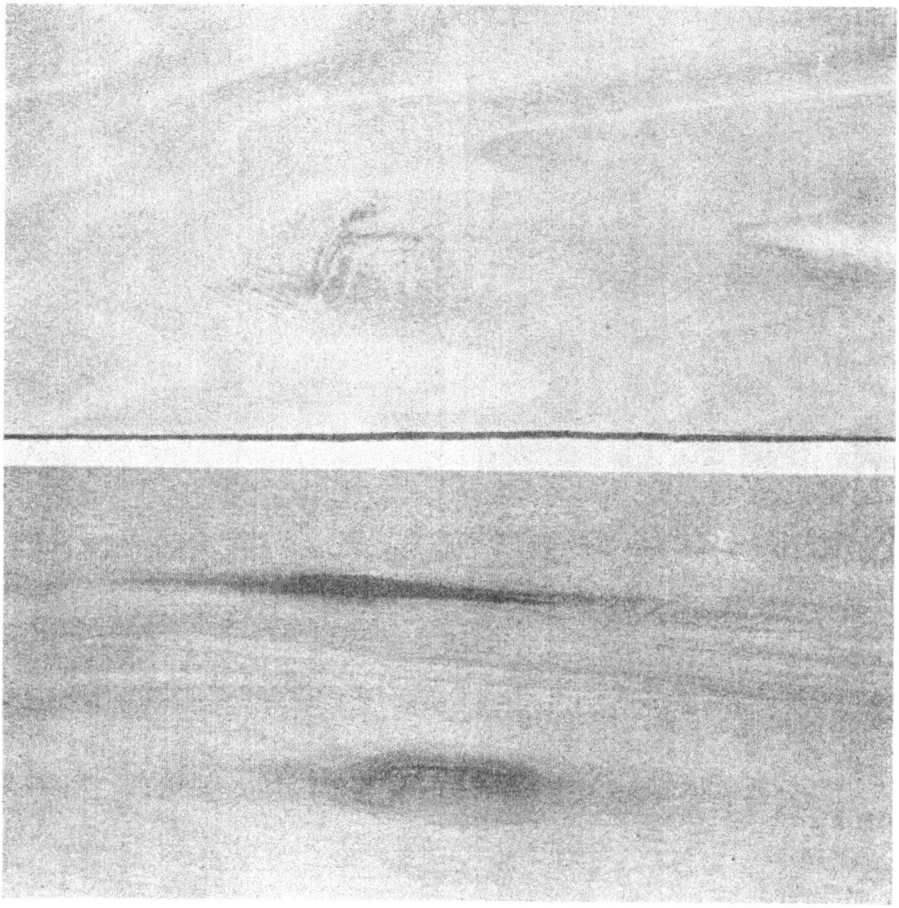

Fig. 43. (Above) Injury to wood during growth. (Below) Bark pockets. (Courtesy Forest Products Laboratory)

streaks are objectionable because they add weight to the wood but not strength. Pitch streaks interfere with proper gluing and finishing.

Stains. There are many kinds of stains and, while not all stains are indications of decay, most decay in its early stages appears as a stain. Stains due to decay often appear as streaks which may extend several feet from the decayed part of the wood. Stains which uniformly discolor annual rings may indicate decay. White or brown stains in wood typically indicate the presence of decay. A reddish-purple color may indicate red heart in its early stages. Soft white pockets are another common

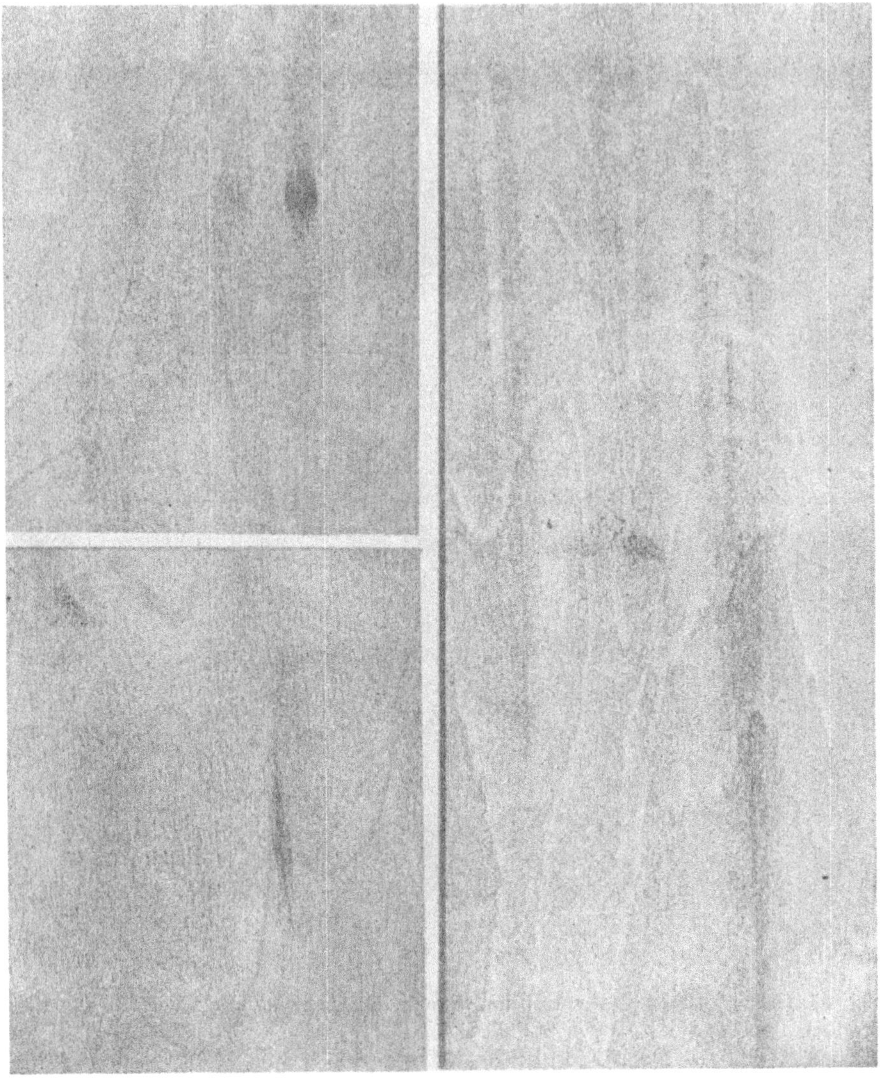

Fig. 44. Pitch streaks. (Courtesy Forest Products Laboratory)

indication of decay. Any wood which is discolored should be examined very closely and not used if it is possible to obtain stain-free wood.

Mineral streaks are defects caused by injury to the wood during its growth. Bird pecks, burls, and lightning cause defects which may or may not be severe enough to cause rejection of the material.

Worm or Insect Holes. When a piece of lumber contains holes caused by worms or insects it should be rejected.

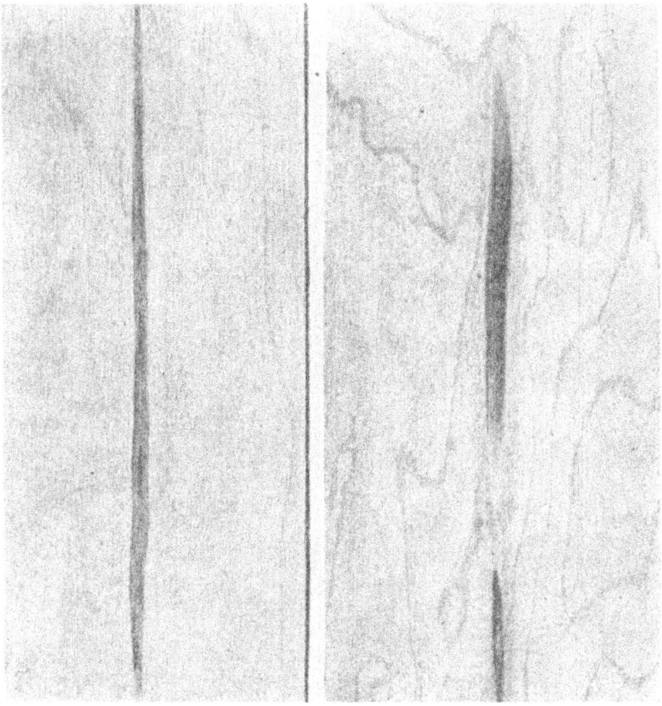

Fig. 45. Mineral streaks in maple. (Courtesy Forest Products Laboratory)

Honeycombing. Honeycombing is a defect which occurs within the wood and is often caused by defective seasoning. Honeycombing is a check developed along the rays of the wood across the annual rings. Figure 46 shows the end of an oak plank which has been honeycombed by too rapid drying.

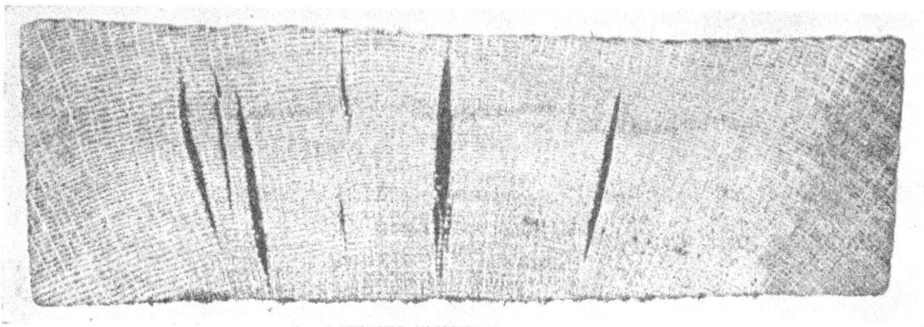

Fig. 46. Honeycombed end of an oak plank. (Courtesy Forest Products Laboratory)

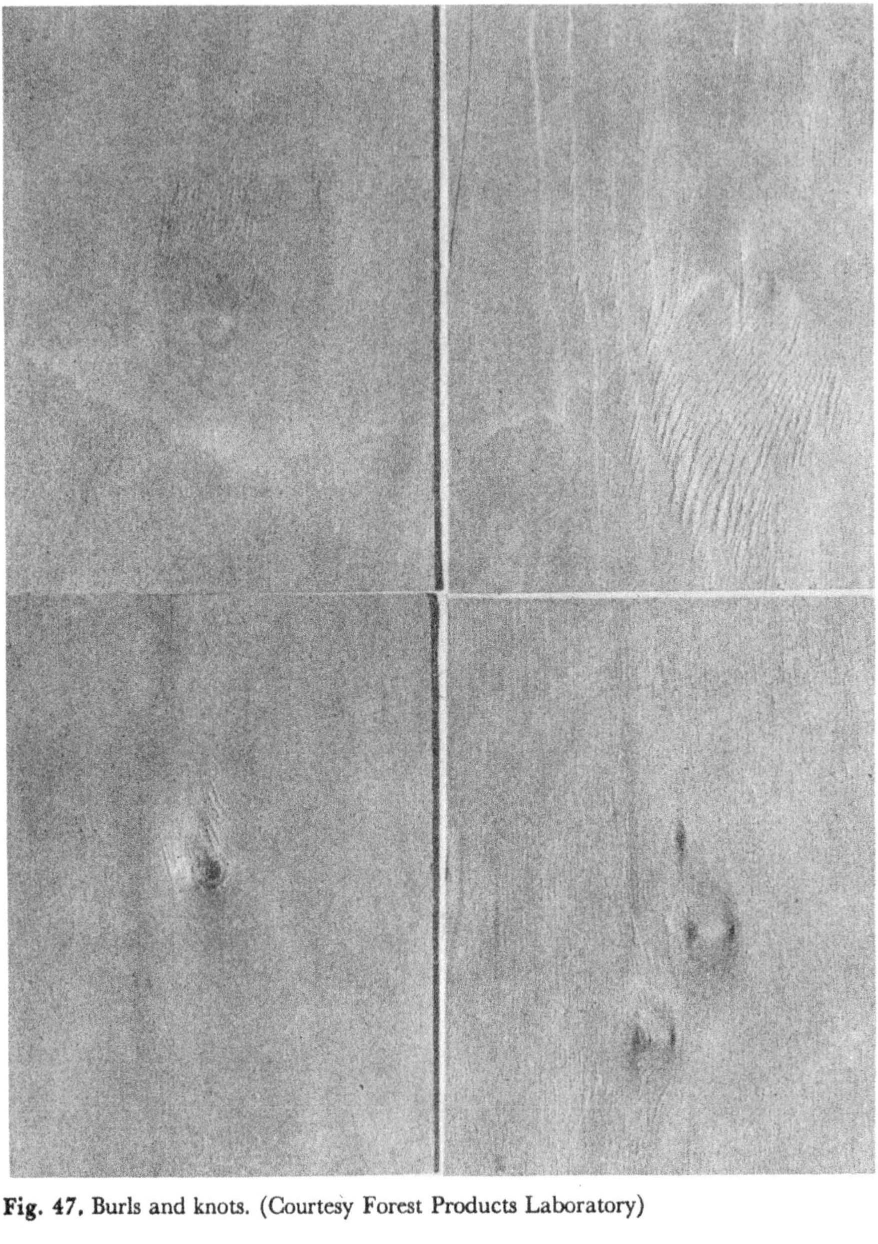

Fig. 47. Burls and knots. (Courtesy Forest Products Laboratory)

Fig. 48. Collapse in various samples of wood. (Courtesy Forest Products Laboratory)

Collapse. Abnormal shrinking may cause grooves to appear in the surface of lumber. If wet lumber is dried at a temperature which is too high, internal stresses are set up in the wood which cause the fibers to collapse. Any indication of collapse is cause for rejection. Red cedar,

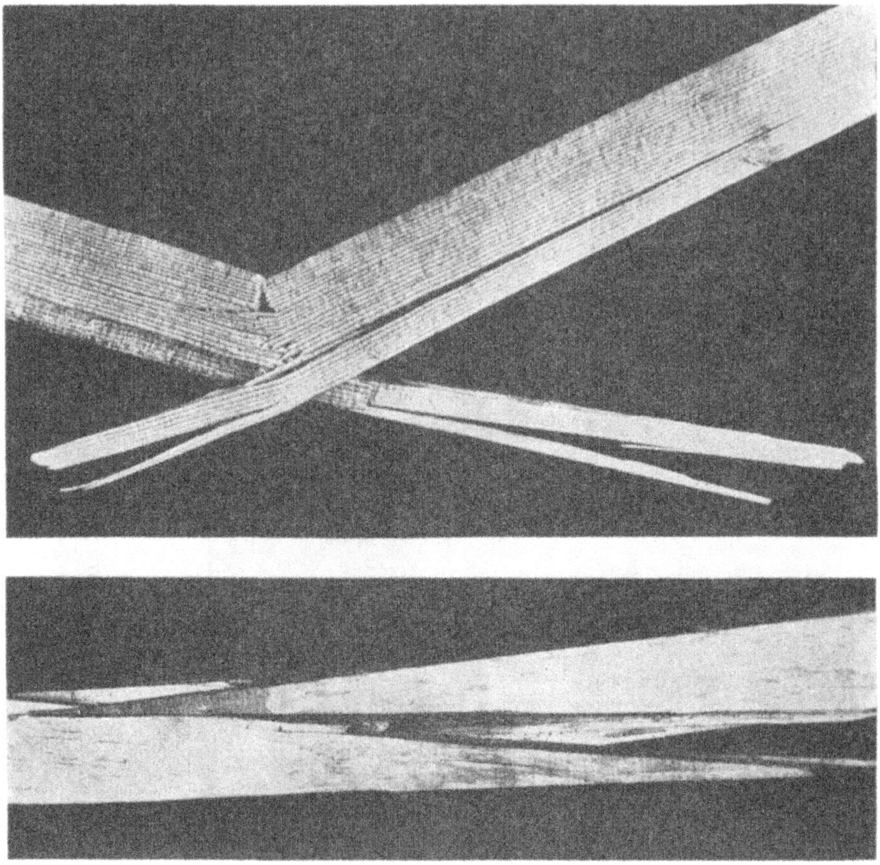

Fig. 49. A typical break of brash wood. (Courtesy Forest Products Laboratory)

redwood, sweetgum, and white oak are particularly subject to this defect. Figure 48 shows the effect of collapse on samples of wood.

Brashness. Brashness is a defect in lumber which is caused by severe temperature changes. Brash wood is usually dark in color and has a lifeless appearance. This wood is brittle, due to sudden shrinking of the fibers. Brashness is cause for rejection. Brash wood breaks easily across the grain without extensive splintering.

Case Hardening. When lumber is kiln dried too rapidly, unequal shrinking causes internal stresses, and case hardening results. One part of the wood may be in compression while other portions are in tension. In the whole piece of lumber these forces balance, and no warping is apparent. When the lumber is sawed or dressed down, these forces are unbalanced, and the material may warp or cup. Figure 50 shows tests for case hardening. Case hardening may or may not be cause for rejection, depending upon the degree permissible for the intended use.

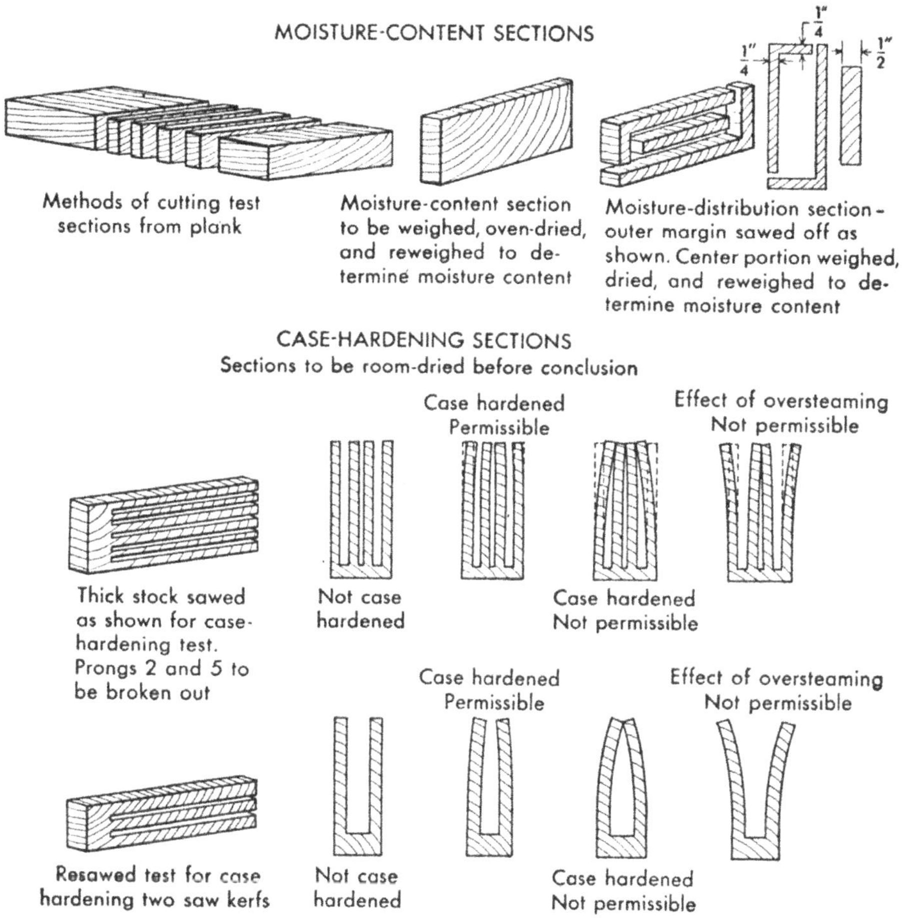

Fig. 50. Test sections for moisture content and case hardening.

V SELECTION OF AIRCRAFT WOOD

It is important to know the requirements of wood to be used in all parts of an aircraft structure. Wood containing imperfections or defects that are not allowable for some parts of the structure may be used in other parts. The selection of any piece of wood depends upon the strength required and the stresses to which it may be subjected while in use. As an example, wood not suitable for spar construction may be used for such parts as glue blocks, filler blocks, stringers, and compression members. This should be kept in mind when selecting wood for any particular part of the aircraft.

General Requirements. All wood used in aircraft structures should, of course, be sound and free from defects which might cause failure of the part. Wood used in parts which are subjected to bending forces should be free from decided cross grain, curly grain, spiral grain, sap pockets, checks, splits, and shakes. Any sign of decay should cause rejection of wood for use in aircraft construction.

Wood for Spars and Spar Parts. All wood used in spars or parts of spars, whether the spar is solid or laminated, should be straight grained and sound. Spars are subjected to higher stresses and carry heavier loads than almost any other part of the aircraft. The slope of grain in spars should never be steeper than 1 in 15. The number of rings per inch for spar stock is given in Table IV.

TABLE IV. LIMITS OF RINGS PER INCH FOR SPAR STOCK

SPECIES	MINIMUM RINGS PER INCH
Cedar, Port Orford	8
Fir, Douglas	8
Fir, Noble	6
Hemlock, western	6
Pine, eastern white	6
Pine, sugar	7
Pine, western white	6
Poplar, yellow	8
Spruce, red	6
Spruce, white	6
Spruce, Sitka	6

Wood used for the making of spars should be within the normal range of specific gravity for that wood. Sapwood may be used if it is bright and sound. Small irregularities in the annual rings are not considered a defect. Wood containing decided discoloration should be rejected. Wood containing compression failures, checks, splits, shakes, or honeycombing should be discarded.

Solid Spars. A solid spar is one which is made from a single piece of wood. The material from which a spar is to be made should be straight

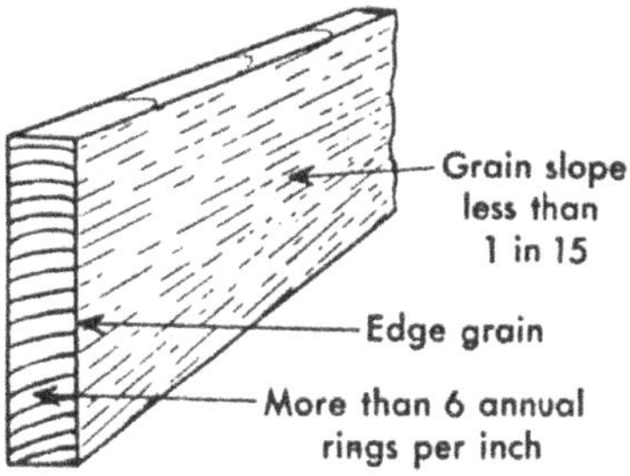

Fig. 51. A solid spar.

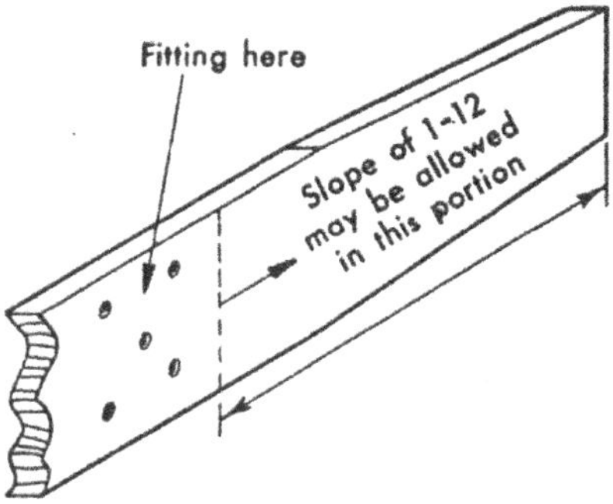

Fig. 52. The tip of a spar.

grained for at least ¾ of its depth. The depth of the spar is its measurement from top to bottom. Edge-grain lumber should be used for making spars. Edge-grain lumber shows less change in volume due to changes in moisture content than does lumber which is flat sawed. Edge-grain lumber has less tendency to cup or warp. The slope of grain for a solid spar should not be steeper than 1 in 15. Small variations in slope which affect less than 5 per cent of the spar may be allowed if the slope is not steeper than 1 in 12. It is sometimes permissible to allow a slope of 1 in 12 near the tip of the spar. Any deviation from the standard minimum in parts of the spar which are near fittings should not be allowed. Spiral grain, curly grain, and interlocked grain should not be allowed unless this defect occurs near the middle part of the depth of the spar and then only for a short length not to exceed a few inches:

Knots having a diameter not greater than ⅛ in. or ⅛ of the width of the face upon which they appear may be permitted if firm and sound. Knots should not be permitted at corners of a spar. It is not good practice to use lumber containing knots for building spars when it is possible to obtain material free from this defect. Knots larger than ⅛ in. are sometimes allowed if they appear entirely within the middle half of the vertical face and do not cause any deviation of grain in the outer quarter of the face.

Pitch pockets in either edge of the spar should not be permitted. Small pitch pockets in the face of the spar which are less than ⅛ in. deep and 1⁄16 in. wide and less than 1 in. in length may be permitted. Pitch

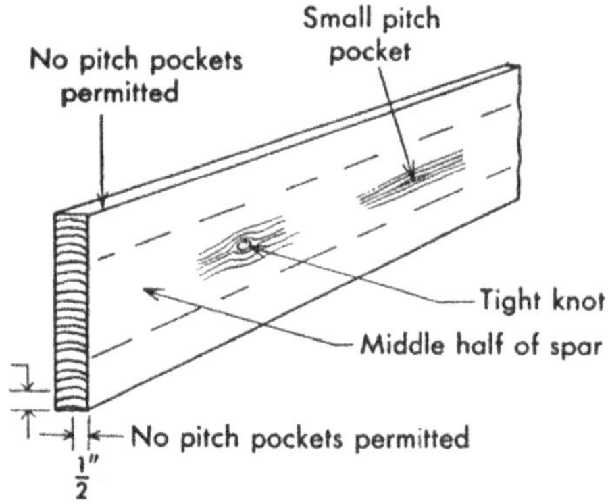

Fig. 53. Pitch pockets and knots.

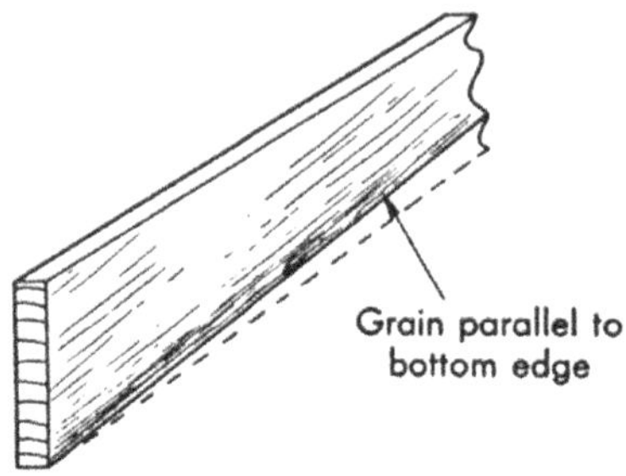

Fig. 54. Grain slope in tapered spar.

pockets of this type are not permitted if they are closer together than approximately 12 in., neither are they allowed if they occur in the same or adjoining annual rings. They should not be allowed closer than approximately ½ in. to any corner of a spar or spar flange. Slightly larger defects of this kind may be allowed if they occur within the middle half of the vertical face of the spar. They are not allowed, however, within the same growth rings or closer longitudinally than about 14 in. The total area of such pitch pockets should not equal more than 2 sq. in. in any square foot of the vertical face of the spar.

Pitch streaks may be permitted on any face of a spar if the streak is less than ½ in. wide. The total width of all of the streaks must be less than ¼ the total width of the spar. No pitch streaks should be allowed within 24 in. of the root end of the spar. Spar material must be free from compression wood or compression failure. When necessary, spar material may be spliced, but only one splice should be made in a spar up to 24 ft. in length. If the spar tapers, the taper should be so arranged that the grain is parallel to the edge of the spar subjected to the greatest stresses. This edge is usually the lower edge of the spar.

Laminated Spars. When a spar is built up by gluing two or more pieces of wood together in such a manner that the glued surfaces are parallel to the face of the spar, the spar is called a vertically laminated spar. The glued surfaces run from top to bottom of the spar, as shown in

Figure 55. In building laminated spars, the grain should be arranged in the laminations so that it runs in different directions in the adjoining pieces. The direction of annual rings should be reversed in alternate layers. Wood selected for vertically laminated spars should meet the same requirements as for solid spars. The same requirements may be

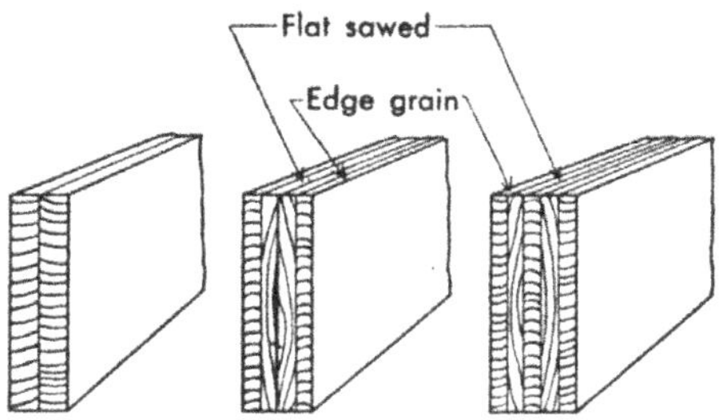

Fig. 55. Vertically laminated spars.

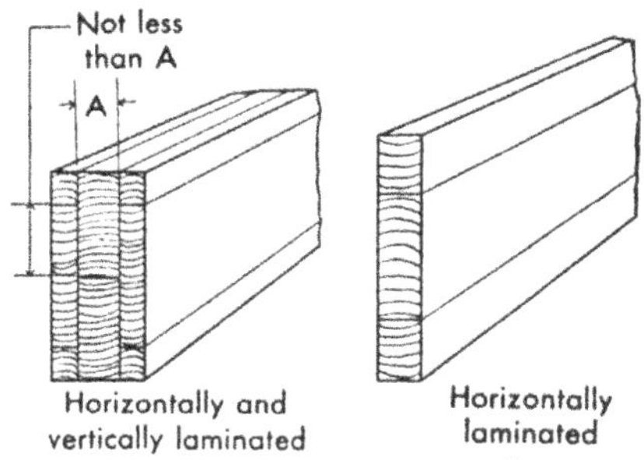

Fig. 56. Horizontally and vertically laminated spars.

used in constructing horizontally laminated spars, the only difference being that the glued surfaces parallel the edges of the spar instead of the faces.

Solid Spar Flanges. A spar flange is formed by gluing a strengthening piece of material along the upper or lower edge of a spar. If this flange is made of a single piece of wood, it is called a solid spar flange. Spar flanges of this type must, if the width exceeds 2½ inches, be edge grain

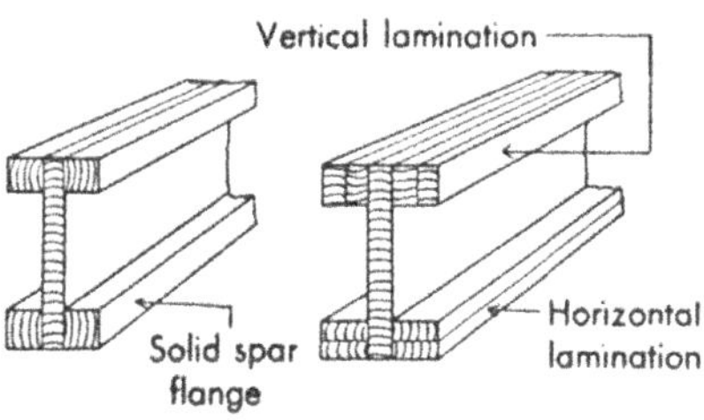

Fig. 57. Solid and laminated spar flanges.

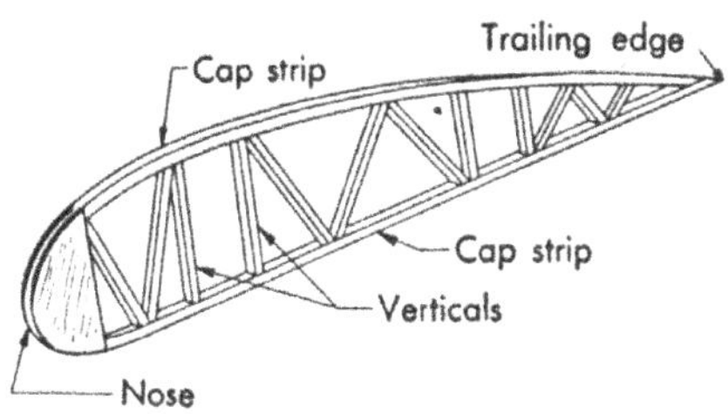

Fig. 58. A wood rib showing parts of a small cross section.

on the horizontal face. Spar flanges of small dimensions may be edge grain on either face. The requirements for material for spar flanges are the same as required for solid spars.

Laminated Spar Flanges. If a spar flange is built up by gluing two or more pieces of wood together, it is called a laminated spar flange. Laminated spar flanges may be formed with the glued surfaces either horizontal or vertical. If laminated flanges exceed 2½ in. in width, the flanges should be edge grain on their horizontal face. (See Figure 57.)

For example, if they are horizontally laminated, edge-grain lumber should be used. If they are vertically laminated, flat-grain lumber should be used. The requirements for wood to be used in vertically laminated flanges are the same as for solid spar material.

Stressed Parts Having Small Cross Sections. Such parts as cap strips and cross members of ribs have small cross sections compared to their length. The highest grade of lumber available should be used in forming these parts. This wood should be entirely free from sap stain, any indication of decay, heart stain, compression failure, compression wood, knots, and pitch pockets. Tiny knots which do not exceed more than ¼ the width of the face may be allowed, provided these knots do not cause a serious deviation of the grain. Curly grain or spiral grain should not be allowed in material used for the making of these parts.

Fuselage Structural Members. Wood used in such fuselage members as bulkheads, stringers, longerons, compression members, and reinforcing blocks should be of high-grade material. If the material is to be bent in forming bulkheads or fuselage-forming members, it should be straight grain and free from knots, checks, cracks, and any indication of decay. If the part is to be glued, it should be free from pitch streaks or pitch pockets. The proper kind of wood should be selected to carry the loads placed upon it in the structure.

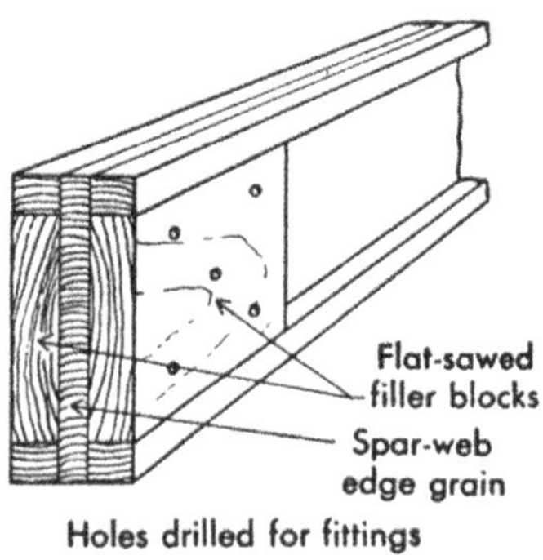

Fig. 59. Filler blocks.

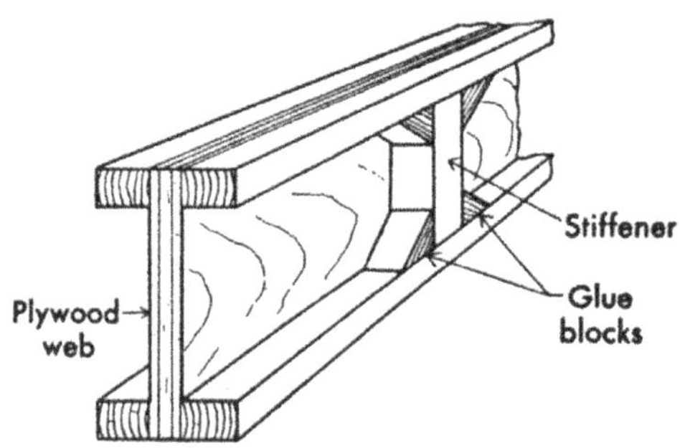

Fig. 60. A plywood web, stiffener, spar flanges, and glue blocks.

Miscellaneous Wood Parts. Lower-grade materials than those previously described may be used for filler blocks, stiffeners, glue blocks, and doorframes. Material for these parts should, however, be free from loose knots, irregular grain, sap streaks, and other defects which are likely to cause splitting where nails and screws are to be used. If these parts are to be glued, pitch stains, pitch pockets, and large knots should be avoided. It is almost impossible to do successful gluing on wood containing large quantities of pitch or resin. Filler blocks should have reasonably straight grain and preferably flat grain on their vertical surfaces.

It is necessary to use considerable judgment in determining whether or not materials for these purposes are suitable. If the material is to be bent and glued to form curved, laminated parts, it should be free from any defects such as cross grain, spiral grain, knots, pitch pockets, pitch stains, or any sign of decay.

VI GLOSSARY OF TERMS USED IN AIRCRAFT WOODWORK

aileron. A hinged or pivoted, movable, auxiliary surface of an airplane, usually part of the trailing edge of the wing, the purpose of which is to maintain the lateral balance of the airplane.

aircraft. Any weight-carrying device or structure designed to be supported by the air, either by buoyance or dynamic action.

airfoil. Any surface, flat or curved, rigid or movable, such as wing, aileron, or rudder, which catches the air in such a manner as to affect the motion of the plane.

airplane. A mechanically-driven aircraft, heavier than air, fitted with fixed wings and supported by the dynamic action of the air.

air screw (propeller). A surface so shaped that its rotation about an axis produces a force (thrust) in the forward direction of the axis.

antidrag wire (T-rod). A wire designed primarily to resist forces acting parallel to the chord of the wing of an airplane and in the same direction of flight.

bark pocket. An opening between annual growth layers that contains bark.

bench hook. A device for holding stock on the bench while it is being worked.

bevel. An angular surface across the edge of a piece of wood.

blemish. Anything, not necessarily a defect, marring the appearance of wood.

brashness. A condition of wood characterized by an abrupt failure across the grain without splintering.

bulkheads. Lateral partitions of an enclosed fuselage.

burl. A large, wartlike excresence on a tree trunk. It contains the dark piths of a large number of buds which rarely develop.

camber. The convex curvature of the surfaces of an airfoil.

cantilever. A kind of wing design so arranged that the entire structural strength of a wing is within the wing itself, requiring no outside bracing.

cell. A general term for the minute units of wood structure.

cellulose. The carbohydrate that is the principal constituent of wood and forms the framework of the cells.

center section. Central panel of the wing structures, usually located directly over the fuselage, to which the wings are attached on each side.

chamfer. To bevel a sharp edge part way down the side.

check. A lengthwise separation of the wood, the greater part of which occurs across the rings of annual growth.

chord. A straight line between the leading and trailing edges of an airfoil.

cloth. Fabric delivered by the bleachery or finisher before it has been proofed, doped, or specially treated for aeronautic use.

cockpit. An open space in an airplane for accommodation of pilots or passengers. When completely closed, it is usually called a cabin.

collapse. The flattening of single cells or rows of cells in heartwood during the drying or pressure treatment of wood, characterized externally by a caved-in or corrugated appearance.

compreg. Plywood, the veneers of which have been treated under pressure and temperature with bonding resin or phenolic resin.

compression failure. Deformations or buckling of the wood fibers resulting from severe stress in compression along the grain.

compression member. A longitudinal strut between spars opposing the pull of wires in the internal drag truss of a wing.

compression rib. A rib which is built particularly strongly with either heavier cap strips or a solid web, or both, for the purpose of withstanding greater compression stress.

compression wood. Abnormal wood that often forms on the lower side of branches and of leaning trunks of softwood trees.

concentric. Circles or spheres which have a common center.

control surfaces. Movable airfoils by means of which the airplane is controlled in flight. They are hinged, auxiliary surfaces and consist of ailerons, elevators, rudders, tabs, and flaps. The control mechanism is the means by which the control surfaces are actuated.

cup. The distortion of a board in which the face is convex or concave transversely.

decay. Disintegration of wood substance through the action of wood-destroying fungi.

defect. Any irregularity occurring in or on wood that may lower its strength.

density. The weight of a body per unit volume.

dope. The finish applied to aircraft fabric, to shrink, preserve, and cause airtightness. Dopes are usually cellulose or collodion compounds, soluble in ether or acetone.

dowels. Small wooden pins used to fasten two pieces of wood in a joint.

drag wire. A wire designed to resist forces acting parallel to the chord of the wing and in the opposite direction of flight.

durability. A general term for permanence or lastingness.

empennage. A collective name for the tail-surface group (rudder, elevators, and stabilizers).

fabric. The cloth (usually a high-grade cotton) used for covering an airplane.

fairing. A member or structure for the purpose of producing a smooth outline and reducing wind resistance.

ferrule. Metal fittings or wire wrappings designed to prevent splitting of wooden parts or loosening of wire terminals, etc.

fiber. A wood fiber is a comparatively long ($\frac{1}{25}$ or less to $\frac{1}{3}$ in.), narrow, tapering cell closed at both ends.

fin. A fixed vertical surface attached to a part of the aircraft in order to secure stability, for example, tail fin, skid fin, etc. Fins are sometimes adjustable.

flitch. A thick piece of lumber with wane (bark) on one or more edges.

former or false rib. An incomplete rib extending from the leading edge to the front spar to assist in maintaining the front curvature of the wing.

framework. The frame of an airplane exclusive of the skin.

fuselage. The structure, approximately streamlined in form, to which are attached the wings and tail units of an airplane.

glue, *casein.* Glue made from the casein which is precipitated from sour-milk curds by the use of hydrochloric acid.

cold-setting resin. An adhesive made from synthetic urea and formaldehyde and containing a catalyst or hardener which affects the setting of the glue at temperatures of 21°C. (70°F.) or above.

hot-setting phenolic type. An adhesive made from phenol or phenolic compounds and formaldehyde, and usually containing a hardener or catalyst. Temperatures of 116° to 160°C. (240° to 320°F.) are usually required to cure these glues.

water- and mold-resistant casein. An adhesive made from casein, lime, and some alkali-producing sodium salt and containing sufficient quantities of a preservative, such as some sodium chlorophenate, to impart mold resistance to the glue joints.

grain. The direction, size, arrangement, appearance, or quality of the fibers in wood.

close-grained wood. Wood with narrow and inconspicuous annual rings.

coarse-grained wood. Wood with wide and conspicuous annual rings.

cross grain. Grain not parallel with the axis of a piece.

diagonal grain. Annual rings at an angle with the axis of a piece as a result of sawing at an angle with the bark of the tree.

edge grain. Edge-grain lumber has been sawed parallel with the pith of the log and approximately at right angles to the growth rings.

flat grain. Flat-grain lumber has been sawed parallel with the pith of the log and approximately tangent to the growth rings.

interlocked-grain wood. Wood in which the fibers are inclined in one direction in a number of rings of annual growth, then gradually reverse and are inclined in an opposite direction.

open-grained wood. Common classification of painters for woods with large pores, also known as "coarse textured."

plain-sawed. Another term for flat grain.

quarter-sawed. Another term for edge grain.

vertical grain. Another term for edge grain.

wavy-grained wood. Wood in which the fibers collectively take the form of waves or undulations.

green. Unseasoned, wet.

gusset. A small connection plate, usually of thin plywood, placed over a joint to give it strength.

hardwoods. The botanical group of trees that are broad leaved. The term has no reference to the actual hardness of the wood.

heart, heartwood. The wood extending from the pith to the sapwood.

honeycomb. Checks, often not visible at the surface, that occur in the interior of a piece.

inspection hole. A small hole having an easily removable cover.

jig. A pattern, form or framework, accurately dimensioned and aligned, in which identical structures of parts can be produced to meet a standard.

joint. A union of two pieces of stock.

joint splice. A joint which consists of two pieces of wood placed end to end and covered with pieces of wood or fishplates bolted or nailed to each side.

kerf. The slot made by the cut of a saw.

kiln. A heated chamber for drying lumber.

knot. That portion of a branch or limb that has become incorporated in the body of a tree.

laminated wood. An assembly built up of plies or laminations of wood that have been joined either with glue or with mechanical fastenings. Distinguished from plywood by the fact that the grain of the wood is in the same direction in all plies.

leading edge. The foremost edge of an airfoil or propeller blade, also called "entering edge."

lightening hole. A hole cut in a solid rib or bulkhead to reduce weight.

lignin. A principal constituent of wood, second in quantity to cellulose. It incrusts the cell walls and cements the cells together.

longeron. A longitudinal member of the framework of an airplane fuselage or nacelle.

loom. The web sections of wing ribs.

lumber. The product of the saw and planing mill not further manufactured than by sawing, resawing, and passing lengthwise through a standard planing machine, crosscut to length and matched.

factory and shop lumber. Lumber intended to be cut up for use in further manufacture. It is graded on the basis of the percentage of the area which will produce a limited number of cuttings of a specified or a given minimum size and quality.

yard lumber. Lumber that is less than 5 in. in thickness and is intended for general building purposes.

boards. Yard lumber less than 2 in. thick, 8 in. or more in width.

dimension. All yard lumber except boards, strips, and timbers; that is, yard lumber 2 in. and less than 5 in. thick, and of any width.

strips. Yard lumber less than 2 in. thick and less than 8 in. wide.

member. Any essential part of a structure or machine.

mineral streak. An accumulation of mineral matter appearing as darkened areas.

moisture content of wood. Weight of the water contained in the wood, expressed in a percentage of the weight of the oven-dried wood.

monocoque fuselage. A type of fuselage construction wherein the structure consists of a thin shell of wood, metal, or other material, supported by ribs, frames, belt frames, or bulkheads, but usually without longitudinal members other than the shell itself. The structure is built so that the shell carries the greater part of the stresses to which the fuselage is subjected.

nacelle. An enclosed shelter for passengers or for an engine. A nacelle is usually shorter than a fuselage and does not carry the tail unit.

nose. The bow of an airplane, usually the front of the fuselage or hull. It also refers to the front end of a wing rib or the leading edge of a wing.

pitch pocket. An opening extending parallel to the annual rings of growth which usually contains, or has contained, pitch.

pith. The small, soft core occurring in the structural center of a log.

ply. A thickness or layer of wood or other substance.

plywood. An assembly made of three or more layers of veneer joined with glue.

rays, wood. Strips of cells extending radially within a tree and varying in height from a few cells in some species to 4 in. or more in oak.

rib, box. A rib built in the form of a solid box at the junction of the wing and fuselage.

rib, form. The lighter, form-giving parts used in airfoil construction.

rib, ordinary. A light, curved, wooden part mounted in a fore-and-aft direction within a surface. The ordinary ribs give the surface its camber, carry the fabric, and transfer the lift from the fabric to the spars.

rings, annual-growth. The growth layer put on in a single growth year.

rotary-cut veneer. (*See* veneer.)

sap. All the fluids in a tree.

sapwood. The layers of wood next to the bark, usually lighter in color than the heartwood, ½ in. to 3 or more in. wide, that are actively involved in the life processes of the tree.

seasoning. Removing moisture from green wood in order to improve its serviceability.

air dried or air seasoned. Dried by exposure to the air, usually in a yard, without artificial heat.

kiln dried. Dried in a kiln with the use of artificial heat.

shake. A separation along the grain, the greater part of which occurs between the rings of annual growth.

softwoods. The botanical group of trees that have needle-like or scale-like leaves and are evergreen for the most part.

spar, main. The front spar within the wing surface and to which all ribs are attached, such spar being the one situated nearest to the center of pressure.

spar, rear. A spar within the wing and to which the ribs are attached, such spar being situated at the rear of the center of pressure and at a greater distance from it than is the main spar. It transfers less than half of the lift from the ribs to the bracing.

specific gravity. The ratio of the weight of a body to the weight of an equal volume of water at some standard temperature.

splice. The adjoining of the ends of two pieces of rope, wire, or wood in such a manner that the splice (joint) is just as strong as the material spliced.

split. A lengthwise separation of the wood, due to the tearing apart of the wood cells.

spring wood. The portion of the annual growth ring that is formed during the early part of the season's growth.

stress. Any internal force acting within a structure.

strut. A compression member of a truss frame — for instance, the vertical members of the wing truss of a biplane (interplane struts) and the shorter vertical and horizontal members separating the longerons in the fuselage.

strut, fuselage. A strut holding the fuselage longerons apart. It should be stated whether top, bottom, or side. If side, then it should be stated whether right or left.

summer wood. The portion of the annual growth ring that is formed during the latter part of the yearly growth period. It is usually denser and stronger mechanically than spring wood.

template. A full-sized pattern from which structural sections are marked out. It may be of paper, cardboard, plywood, or metal.

trailing edge. Rearmost edge of an airfoil or of a propeller blade.

veneer. Thin sheets of wood.

rotary-cut veneer. Veneer cut in a continuous strip by rotating a log against the edge of a knife in a lathe.

sawed veneer. Veneer produced by sawing.

sliced veneer. Veneer that is sliced off by moving a log, bolt, or flitch against a large knife.

wane. Bark, or lack of wood from any cause, on edge or corner of a piece.

warp. Any variation from a true or plane surface.

wing. The main lifting and supporting surface of the airplane in flight, designated as right and left. The right and left sides of an airplane are relative to the right and left hand of the pilot seated in the cockpit.

wing rib. Chordwise member of the wing structure of an airplane used to give wing-section form and to transmit load from fabric to spars.

wing tip. Outer or outboard end of a wing.

workability. The degree of ease and smoothness of cut obtainable with hand or machine tools.

VII PLYWOOD

Plywood, as understood in the modern sense, dates from shortly after the Civil War. The process of veneering goes back to ancient Egyptian times.

Plywood always has an odd number of layers. Simple plywood is made from three thin layers of wood glued together with the grain of the middle layer at right angles to the grain of the outer layers. Wood, when it absorbs moisture, swells across the grain. The placing of the inner layer of plywood with the grain at right angles to the outer layer reduces this swelling considerably. This is largely due to the binding effect of the glue. Usually the inner layer is equal to approximately twice the thickness of the outer layers, to overcome somewhat the tendency of the layers to separate because of swelling and shrinking due to the absorption of moisture. A piece of wood as thin as the average layer of plywood would split very easily. The crossing of the grain in plywood tremendously strengthens the layers. Each layer of plywood is called a "ply" or "veneer."

Much modern furniture is built up from comparatively cheap wood and covered with a thin veneer of more expensive and more attractive wood.

Wood has most of its strength in one direction. Most wood is weak across the grain and splits easily. Plywood does not tend to split when nails are driven into it or when placed under stresses.

The dimensions of plywood in a structure tend to remain the same, due to its slight tendency to shrink or swell. Plywood sheets can be made in almost any size. Plywood, due to the fact that it comes in comparatively thin layers, can be steamed or wet and bent into almost any shape.

Most plywood is glued with synthetic resin glues under conditions of heat and pressure which make the wood layers practically a single unit. Plywood which is ¼ in. thick is much stronger if built from layers

1/32 in. thick than if built from layers 1/16 in. thick. The more layers per unit thickness, the greater the strength.

Some of the modern plywoods are bound together by plastics which are forced completely through the layers of wood under conditions of

Fig. 61. Logs to be used in the manufacture of plywood. Notice the heartwood and sapwood on the ends of the logs on the car. (Courtesy Curtiss-Wright Corporation)

heat and pressure. This forms a sheet which is stronger than ordinary plywood.

A new process has been perfected whereby "plymetal" is fabricated. Plymetal consists of a thin sheet of metal between two thin sheets of plywood. This material is very strong and can be used as lining for baggage compartments and covering for floors in the cabins of transport airplanes.

Most of the plies, or veneers, are cut on a machine which closely resembles a lathe. This is called rotary cutting. The cutting knives shave off the thin layers at high speed. The speed of cutting is rapid enough so that the fibers of the wood are not greatly torn or disturbed. The pieces of log or, as they are commonly called, "bolts" are boiled in water before cutting. The material remains in the boiling water until thoroughly saturated. The bark is removed after "cooking," and the bolt placed on the cutting machine while the material is still hot and wet.

Fig. 62. Rotary cutting of plywood veneers. (Courtesy Curtiss-Wright Corporation)

Some veneers are cut by slicing on a slicing machine. The slicing method has the advantage that the sheets are cut flat, and the grain is not stretched on one side and compressed on the other, as is the case when the veneer is cut on a rotary lathe. Rotary-cut veneer always has an open and a tight side. The tight side of the surface layers of plywood should be to the outside.

Sheets of plywood may be joined together by means of a scarf joint and glued so that it is possible to have a sheet of any size and shape desired.

Forming of Plywood. Most plywood for use in the aircraft industry is manufactured and furnished to the aircraft builder in the form of flat sheets. These sheets must be bent or molded to the required form by the aircraft worker. The amount that the sheet can be bent is influenced by several conditions, such as thickness, moisture content, grain of the plywood, temperature, quality of the plywood, and the method used in bending. Much plywood can be formed dry without any special treatment. Special treatment consists of soaking, steaming, or placing

Fig. 63. Cutting flitches for the plywood-slicing machine. (Courtesy Curtiss-Wright Corporation)

in hot or boiling water. Where soaking or steaming is applied, care must be taken to be sure that the glues used in forming the plywood will not be affected by the added moisture. Each kind of plywood in each thickness has a very definite bending radius beyond which breaking will take place. (See Table V.)

Pieces of plywood used in aircraft construction should not be subjected to the maximum bending as this develops too much strain and may cause failure. Plywood can be bent more abruptly if the grain of the surface layers is placed parallel to the curve around which the wood is being bent. When plywood is dried after being wet, there is always a tendency for it to flatten out to its original form. Wherever possible, plywood should be formed to the shape required before placing in the structure.

A piece of plywood 1/64 in. thick can be bent around a curve with a radius of less than 1/2 in. Where plywood is used over double curved surfaces, or irregular curved surfaces, the problem becomes an inde-

Fig. 64. Rough-cut plywood. (Courtesy Forest Products Laboratory)

TABLE V. APPROXIMATE MINIMUM RECOMMENDED BEND RADII FOR AIRCRAFT PLYWOOD

PLYWOOD THICKNESS		10 PER CENT MOISTURE CONTENT, BENT ON COLD FORMS		THOROUGHLY SOAKED IN HOT WATER AND BENT ON COLD FORMS	
		Bend at Right Angles to Face Grain of Plywood	*Bend at 0° or 45° to Face Grain of Plywood*	*Bend at Right Angles to Face Grain of Plywood*	*Bend at 0° or 45° to Face Grain of Plywood*
inch	no. plies	inches	inches	inches	inches
1/32	3	2.0	1.1	0.5	0.1
1/16	3	5.2	3.2	1.5	0.4
1/8	3	12.0	7.1	3.8	1.2
5/32	3	16.0	10.0	5.3	1.8
3/16	3	20.0	13.0	7.1	2.6
7/32	5	27.0	17.0	10.0	4.0
1/4	5	31.0	20.0	12.0	5.0
5/16	5	43.0	28.0	16.0	7.0
3/8	5	54.0	36.0	21.0	10.0

pendent one and must be solved by the woodworker. He must be able to decide whether it is necessary to soak or steam the plywood and, in extreme cases, he may have to build a form or die with which to form the plywood before attaching it to the structure. The development of

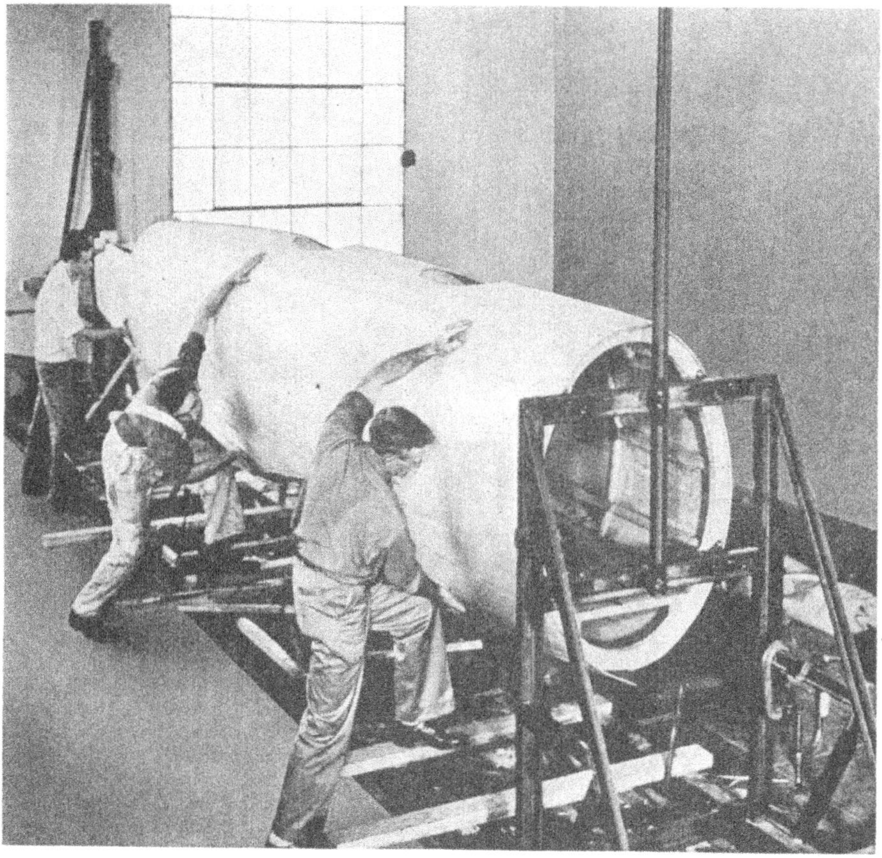

Fig. 65. Assembling a molded plywood fuselage. (Courtesy Timm Aircraft Corporation)

the synthetic resin glues has made the process of soaking and steaming plywood much more practical in recent years.

In selecting plywood, smoothness is of importance. Some plywood is rough cut and this roughness shows up on the surface of the plywood, causing thin spots which sometimes can be detected by holding the plywood up to the light. Whenever possible, select sheets of plywood which have dried flat, avoiding any which have a wavy appearance. These sheets are very difficult to press into shape without special equipment. The same defects may occur in plywood as occur in other wood and have the same effect. Each sheet of plywood should be carefully

Fig. 66. Wing construction showing a molded plywood leading edge, gussets, and rib webs. (Courtesy Universal Moulded Products Corporation)

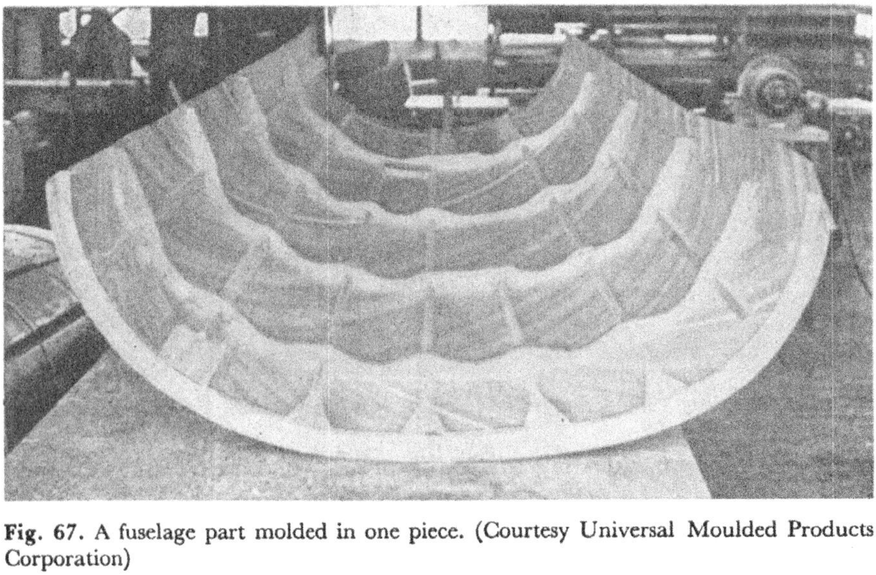

Fig. 67. A fuselage part molded in one piece. (Courtesy Universal Moulded Products Corporation)

Fig. 68. Assembling a molded fuselage. (Courtesy Universal Moulded Products Corporation)

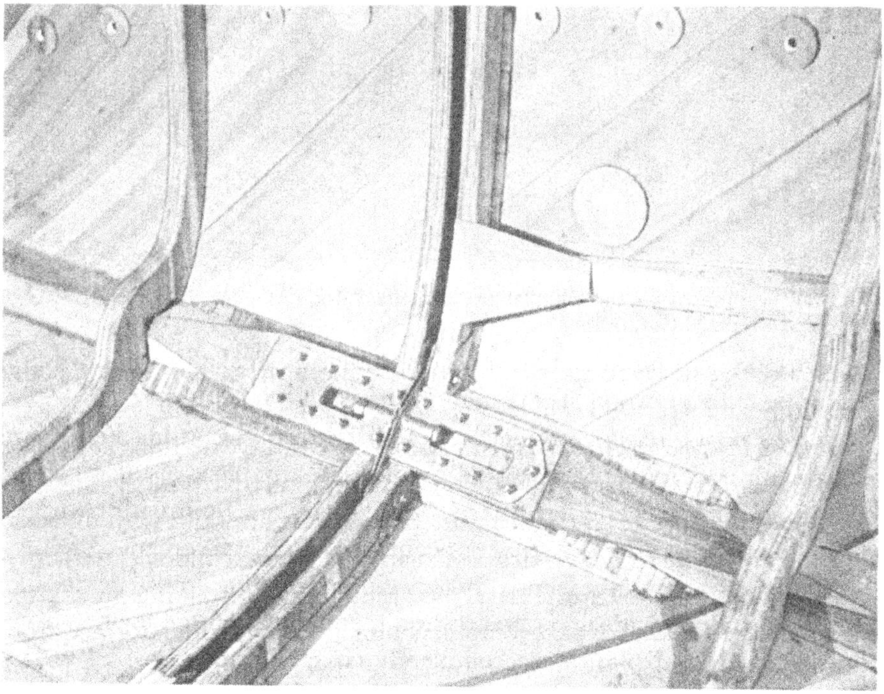

Fig. 69. The method of joining the molded parts of a fuselage. (Courtesy Universal Moulded Products Corporation)

examined before being placed in an aircraft structure. Very few sheets of plywood are entirely free from some form of imperfection. The woodworker must use good judgment in determining whether a piece of plywood has defects sufficient to cause its rejection. Small, scattered, tight knots are quite common in mahogany and walnut veneers and are not cause for rejection. When a piece of plywood is used to reinforce a spar where fittings are to be attached, no visible defects should be permitted.

Fig. 70. Equipment for the manufacture of molded wood products. (Courtesy Universal Moulded Products Corporation)

Pitch streaks, sap pockets, bark pockets, mineral streaks, rots, and stains may appear in plywood sheets and are cause for rejection.

Uses of Plywood in Aircraft Structures. The use of plywood has kept pace with the aircraft industry. Plywood is used in many parts of the airplane for reinforcing joints and light parts. Many spars are built with plywood webs. Points where heavy stresses occur are often reinforced with plywood. Plywood is used extensively for reinforcing glue joints in rib and spar repairs. Bulkheads and compression members are often formed from plywood. Many airplanes are completely covered with plywood in the form of stressed skin. It is good practice to reinforce the leading edge of fabric-covered wings with plywood.

Molded Plywood and Structures. The molding of plywood into complicated forms and curves has developed rapidly within the last few years, particularly in the aircraft industry. By making use of the synthetic resin glues, plywood layers which have been impregnated with the glue can be molded into any shape for which a mold can be made that will allow the piece to be separated from the mold when finished. Almost any part of an airplane can be made by this method. Not only the ply-

Fig. 71. The manufacture of small bonded plywood sheets. (Courtesy Curtiss-Wright Corporation)

wood skin but structural members can be fabricated in a single operation by premolding. Pressure is applied in the molds by means of rubber bags, waterproof canvas forms, wood molds, steel molds, and molds of concrete. Bags of various materials may be used. Some of these bags, such as those formed from cellophane, can be used only once. Heat is furnished by means of steam, hot water, ovens, and electronic waves.

Fig. 72. Removing the rubber bag from the fuselage mold. (Courtesy Timm Aircraft Corporation)

VIII FUNDAMENTALS OF AIRCRAFT WOODWORKING

Aircraft woodwork requires a high degree of skill and careful workmanship. All parts of an aircraft structure must be made to exact size and fit. There are certain fundamental rules which must be followed in the working of wood. The woodworker must work to exact size as shown by the working drawings and must so plan his work that too much material is not removed, as material once removed cannot be replaced.

In laying out his work, the lines with which he works should be as fine as possible. Either a very sharp pencil or a sharp-pointed implement should be used for marking off dimensions. If a sharp-pointed instrument is used, the mark should be so placed that it will be removed and not be left on the finished part.

In using a saw, the cut should be made so that it comes exactly to the line. The line itself should not be used as a center guide for the cut. A saw removes a layer of wood equal to the greatest width of the teeth of the saw. A common fault with beginners is to leave the part being sawed off unsupported. The weight of the unsupported part will cause breaking of a portion of the unsawed material before the saw has finished its cut. This usually causes splintering of the material, making it unfit for use. Use as fine a saw as possible for the job at hand.

In using planes for smoothing wood surfaces, care should be taken not to plane against the grain, as this tears loose small splinters of wood, leaving a rough surface, as shown in Figure 73. Care should be used not to split small pieces of wood by driving of nails or placing undue strain on small parts.

Plywood should be cut with a knife or very fine saw.

A miter box should be used whenever possible in cutting joints that require close fitting. It should also be used in cutting small parts such as cap strips and rib diagonals.

Hammer marks and bruises of various kinds are easy to place in wood but difficult to remove.

Most wood parts of an aircraft do not require fine finishing. Surfaces prepared for gluing should not be finished to a high polish. Most of the woodwork in an aircraft is not exposed to view. Exposed surfaces are usually covered with a finish which conceals small imperfections.

Fig. 73. Effect of planing against the grain of the wood. (Courtesy Forest Products Laboratory)

Wood should not be fed too rapidly into any kind of power woodworking machine. It is better to take two or three thin cuts with a jointer than to take off a thick cut in one operation. Material should not be forced too rapidly through power saws. Circular saws should be fed at a reasonable rate. Power machinery tends to get away from the operator, and much fine material has been spoiled by careless use of power machinery.

The drilling of holes for bolts to hold fittings in place requires great care and a considerable degree of skill. Jigs should be used wherever possible, particularly if slanting holes are drilled. Most beginners place too much pressure on the drill when boring through wood. The wood may be splintered on the opposite side if a suitable backing block is not used. To avoid splintering, counterboring may be used. When the hole is drilled part of the way through from one side and then finished by boring from the opposite side, this procedure is called **counterboring.**

Wood is easily worked but requires careful workmanship. Wood should not be worked unless the moisture content is within the specified limits. Often, inferior wood is discovered by the worker and may have to be discarded even after considerable time and effort have been placed upon the part being constructed. It is better to discard material which has been slightly damaged than to risk the collapse of the whole structure. Under no circumstances should the worker in aircraft construction conceal a defect.

The workman should always use the proper tools for the job. Tools should never be used for any purpose other than that for which they were designed.

Bent Work. Where it is desirable to use solid wood and it is necessary that the wood be bent, straight-grained, clear wood of the best quality should be chosen. Oak, spruce, and ash are the woods which are most commonly used when it is necessary that the part be bent. Only hardwoods are recommended for bending. Most softwoods, because of high pitch content, are not suitable for bending when steaming or soaking is required. To make simple bends, soaking the wood in water is all that is necessary. When a severe bend is to be made, steaming is recommended. Steaming for more than 4 hr., or boiling for more than 6 hr., is apt to injure the wood and destroy a considerable part of its strength. After wood has been soaked and bent into shape, it should be left in the form until thoroughly air dried. In steaming wood, a simple steam chamber may be constructed.

Plywood is seldom applied to aircraft structures without some bending. Many bends are made having relatively short radii, such as over the leading edge of a wing and the leading edge of streamlined parts. Most plywood will stand a considerable degree of bending. It is often necessary that plywood be soaked in water. Most plywoods are prepared with waterproof glue, and soaking does not injure the glued plywood. Where the radius is comparatively small it may be necessary to treat the plywood with hot water or steam. It should not be steamed for more than one hour or boiled for more than two hours. Most plywoods can be bent over the required forms with but a few minutes of steaming or soaking in boiling water. After softening in water, plywood may be bent over a form and allowed to dry, or bent directly over the part to which it is to be fastened. Plywood should be allowed thoroughly to air dry in shape before gluing or fastening. A piece of burlap placed over the wet plywood assists in rapid drying. While drying, plywood should be where there is a good circulation of air. When a good many pieces of the same shape are required, they may be placed in a single form, but this lengthens the time of drying. Plywood, in drying, may be heated to as much as 300°F. without danger of damage to the plywood. Plywood should be removed from the form as soon as thoroughly dry.

Fig. 74. Nail-strip gluing. (Courtesy Forest Products Laboratory)

Nail-Strip Gluing. It is sometimes difficult to obtain the proper pressure to glue plywood to the structural parts. Pressure may be obtained by the use of nailing strips. Figure 74 shows how this can be performed. Nailing strips are usually $3/16$ in. thick and $1/2$ in. wide. They are cut from soft-textured woods which do not split easily, such as white pine or basswood. The nails should be spaced not more than 1 in. apart, and if more than one row of nails is used, the nails should be staggered in the alternate rows. Nails should extend through the plywood not less than 3 times the thickness of the plywood. No. 18 or No. 20, treated, flat-head, steel nails of the proper length should be used. Nail-strip gluing is not recommended when the plywood is thicker than $3/16$ in. if the plywood is used as a skin covering for a wing or control surface. It is better practice to drive the nails with several light blows than with a single heavy blow. Nailheads should be flush with the surface of the nailing strip. Care should be taken not to injure the skin. In removing the nailing strip, a pair of pincers is the most desirable tool, as a claw hammer or other prying tool is almost sure to damage the skin. A strip of tape may be placed under the nailing strip, which will assist in its

Fig. 75. Nail-gluing ribs showing the use of a jig. (Courtesy Forest Products Laboratory)

removal. It is recommended that a heavy layer of paper be placed under the nailing strip extending approximately ½ in. from under the strip on each side. This helps to prevent damage to the skin.

Nail Gluing. Nails are sometimes used for applying pressure. The nails are driven in such a manner that they hold the parts being glued together with sufficient pressure to assure a good glued joint. The nails used in this method are not removed as in nail-strip gluing, therefore they should be of a noncorroding type. In plywood ⅛ in. or less in thickness, the nails should be spaced approximately ½ in. apart. For thicker plywood, spacing of 1 in. may be used. Nail gluing is not recommended for plywood more than ¼ in. thick. The nails should extend through the plywood a minimum of three times the thickness of the plywood. Nail gluing is not usually used in applying skins, as nails have a tendency to work loose and crack the finish or produce an uneven surface. If the plywood is fabric covered, the nails have a tendency to stain or rot the fabric. Rib gussets and gussets applied to other internal structures are commonly nail glued. Nail gluing is often used in applying plywood webs to ribs, spars, etc. It is also commonly used in gluing flanges or strips used to reinforce spars and other similar structures.

Bolts and Bushing Holes. It is necessary that all bolts and bushings used in aircraft structures fit snugly into the holes. Any looseness in the hole allows the bolt or fitting to work back and forth, and the hole soon becomes enlarged. Rough holes are often caused by using dull bits or trying to bore the hole too rapidly. Well-sharpened twist drills produce smooth holes in both solid wood and plywood. The twist drill should be sharpened to approximately a 60° angle. A steeper angle, approximately 118°, should be used for high-speed drills. Whatever the speed of the drill, it should be fed only rapidly enough to cut thin shavings. The drill should not be pressed into the wood fast enough to cause "chattering." High-speed drills may "sing" but should not "chatter." When drilling by machinery, the wood should always be clamped to the drill table so that no movement can take place while the hole is being drilled. Holes to receive bolts should be of such size that the bolt can be inserted by light tapping with a wood or rawhide mallet. If the hole is so tight that heavy blows are necessary to drive the bolt, deformation of the wood takes place which may cause splitting or unequal load distribution. All holes bored for bolts which are to hold fittings in place should match exactly the holes in the fitting. Even the smallest fraction of an inch out of alignment will cause unequal load distribution. The fitting itself should not be used as a jig. A jig should be prepared from a piece of hardwood, or, if a metal jig is used, the metal should be soft so as not to injure the drill itself. Bushings are often made of plastic or light metal which provides additional bearing surface without any great increase in weight. Sometimes light steel bushings are used to prevent crushing the wood when bolts are tightened. The holes for bushings should be of such size that the bushing may be inserted by means of light tapping with a wood or rawhide mallet.

IX TECHNICAL DRAWINGS AND THEIR INTERPRETATION

It is becoming more and more important that the worker in wood in aircraft construction be able to interpret engineering drawings rapidly and accurately. It is by means of this type of drawing that the workman can exactly duplicate the ideas of the designer. The workman usually receives a copy of the engineering drawings or the part of the original drawings representing the part which he is to build. These drawings are called "working drawings."

Most work is done from blueprints in the form of detailed drawings. There are a number of different ways in which these drawings are furnished to the workman. The blueprint is the most common. On blueprints the entire background is blue, and lines representing the drawing, dimensions, and figures appear in white.

There is also the brownprint which is not different from the blueprint except that the background is brown instead of blue. The lines on this drawing appear in white against a brown background. Blúeprints and brownprints are made directly from the tracing, which is a duplication in ink of the drawing on translucent tracing cloth. Drawings are printed from the tracing cloth in much the same manner that a picture is printed from a negative on sensitive paper to make the finished photograph.

A blue-lined print is a print upon which the lines appear in blue against a white background. There are also brown-line and black-line prints. By means of photography, photostats are obtained, which may be white lines on a black background which is the negative, or black lines on a white background which is the positive. The latter method is used to enlarge or reduce drawings.

Pictorial Views. Pictorial views picture the object in such a manner that several sides of the object are seen. Usually the sides which appear are the right side or face, the plan or top face, and the front face or front view. Such a view gives one the impression of seeing the whole object. On this type of drawing the lines are foreshortened when drawn

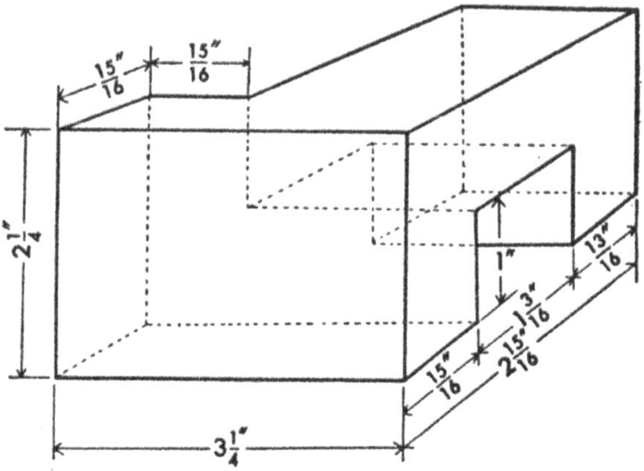

Fig. 76. A pictorial drawing with dimensions.

in perspective in order to make the object appear to have three dimensions. In engineering drawings the three faces which appear in the view are drawn as though viewed at right angles to the face. For example, a rectangular block, as shown in Figure 77, viewed directly from the front appears as a rectangle. This is called the front view. Viewed from the right side, another rectangle is seen which is drawn to the right of the front view and is called the side view. Directly above the front view is drawn the third rectangle which is called the top or plan view. In making these three drawings, it is customary to place the front view in the lower lefthand part of the page with the top view directly above it and the right-side view to the right of the front view. These three views are commonly called a "three-view drawing."

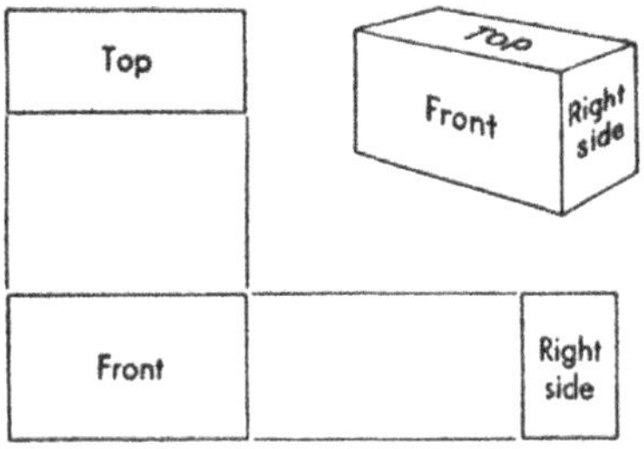

Fig. 77. Three-view drawing of a rectangular block.

Blueprint Lines. Lines representing the edge of an object which can be seen are drawn heavily. These are called object lines. To represent an edge which is hidden behind some other part of the object, dashed lines are used. These lines are called invisible object lines. Sometimes a heavy object line on a view may represent both the visible

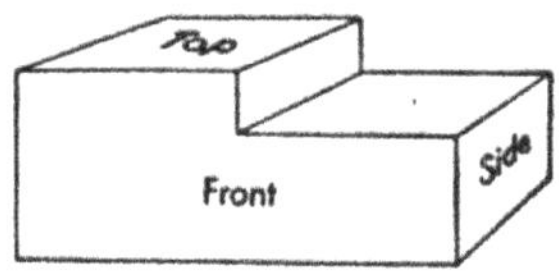

Fig. 78. Pictorial drawing of an object.

and the invisible lines, when the invisible line is directly behind the visible line.

The beginner usually has some trouble in getting a picture from these three flat views. With practice, however, he can, from these three views,

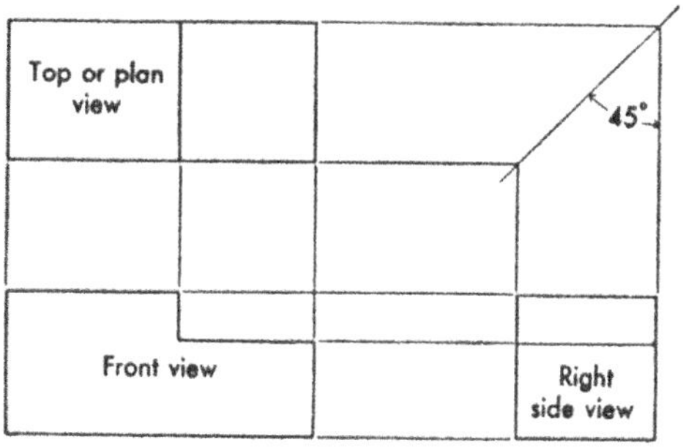

Fig. 79. Three-view drawing of the object shown in Fig. 78.

get a mental picture of the complete object. As he gains experience he can visualize a very complicated object from the three views shown in this type of drawing.

The corresponding parts of each view are usually connected by thin solid lines which are called extension lines. Dimension lines show the

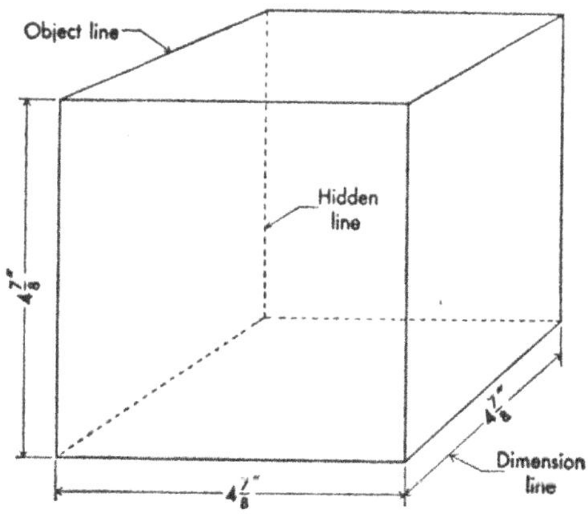

Fig. 80. A pictorial drawing showing object lines and hidden lines.

distance between various parts of the object illustrated. Between these lines are placed a broken arrow and the dimensions in inches or feet. Dimension lines are illustrated on the drawings used in Figures 76 and 80.

Sections. Sectional views are made by cutting through a part of the object and drawing the area which has been cut. These views are often used to show important details which cannot be otherwise shown on the drawing. Usually a line is made on the front, side, or top view showing where the cut was made. Letters are used to indicate where the cut was made. On the sectional drawing the same letters are used. A sectional view is usually marked with lines which represent the type of material cut, which may be wood, steel, brass, etc. These cross lines or markings are called conventional, cross-sectional markings.

A complete drawing or assembly drawing may represent either the complete structure which is being made, as for example an airplane, or

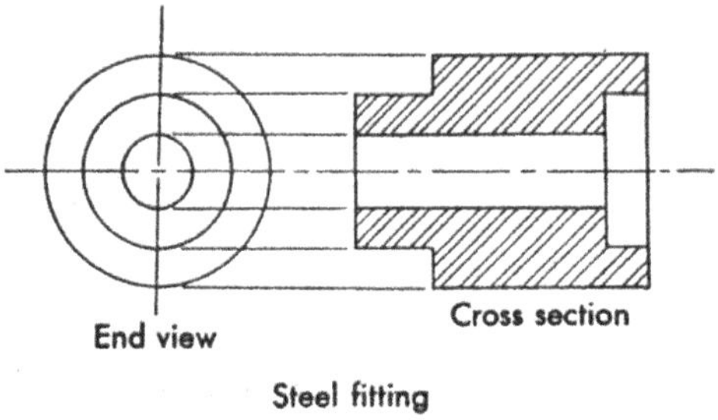

Fig. 81. A typical cross section.

a part of a complete structure, such as an airplane wing, or a drawing showing how a smaller fitting is to be made. A detailed drawing shows only one small part in detail. Detailed drawings give all information needed, even to the kind of nails, size, and number to be used for each fastening. Each detail is plainly shown on the drawing to enable the workman to complete the part. It would be almost impossible to build an airplane from a complete assembled drawing. Each part is broken down into detailed drawings which usually make it possible for one workman to complete all the work on the part shown on his detailed drawing. Broken sections are used where it would be impossible to show a very long part on the drawing. For example, a spar may have its root and its tip very close together on the drawing, with the greater portion of the spar left out. This is made clear in Figure 82. All blueprints and technical drawings follow the same pattern in order that any person familiar with this type of drawing will know exactly where to look to obtain needed information. In the lower righthand corner of the drawing is a box containing the title of the drawing in large letters. This title gives the name of the assembly and the name of the job. On a small detailed drawing, this title may simply give the name of the operation to be performed or the name of a single part to be made. Other information needed to clarify or make plain the title is often placed in this box or in ruled-off portions of the box. The number of the drawing is placed to the right of the title. In this box will be found the name of the engineer who made the drawing or the company producing the part. The name of the draftsmen may appear. If the drawing has been checked by another engineer or draftsman, the name of this person will also be placed in this box. The date of the drawing and the date approved are noted.

All materials used on the job are listed in a box near or above the title box. This box may have the word, "noted," in it which indicates

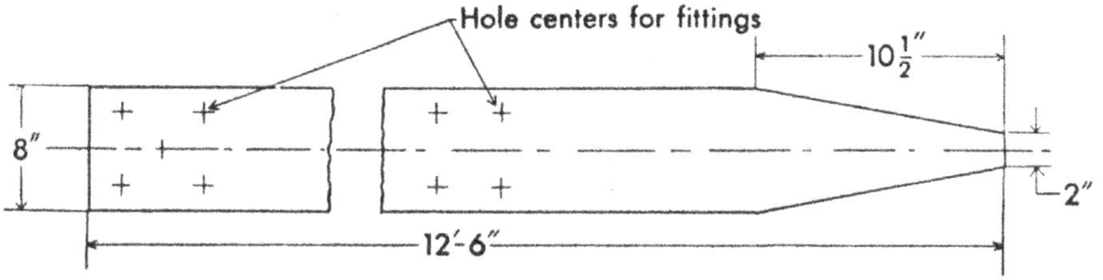

Fig. 82. A broken view showing the root and tip of a spar.

that notes pertaining to the proper material are found elsewhere on the print. On blueprints the various parts are numbered. This enables the workman to get a detailed blueprint or drawing of that part.

Specifications are directions which accompany drawings describing the kinds of material to be used. The driving of an extra nail or the placing of an extra screw might weaken instead of strengthen the part. Each workman should have a good understanding of blueprints and drawings. A person learns to read a blueprint much as he learns to read the printed page. A blueprint which looks very complicated to the beginner is as readable to a person familiar with the symbols and lines as is the printed matter in a simple school book.

X GLUES AND GLUING

No wood aircraft structure or repair is any stronger than its gluing. Glue makes possible the use of woods in the construction, maintenance, and repair of aircraft structures. The proper glue bond holds wood structures together more tightly than any other means yet devised. The glue adds a minimum amount of weight in the joint formed, and its adhesive action is generally stronger than the wood joined. The adhesive action of a glued joint is due to the glue's being forced into the pores of the wood

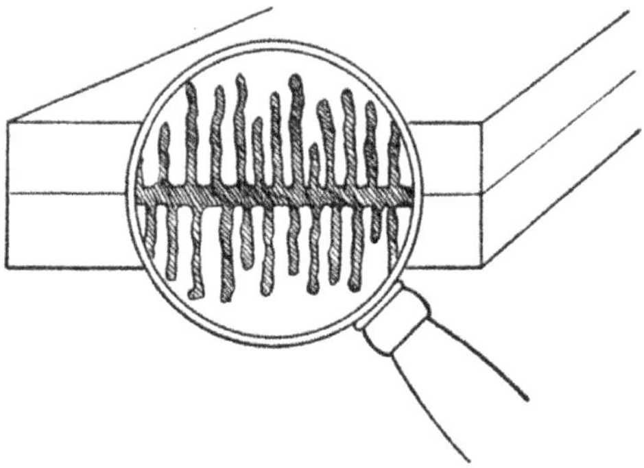

Fig. 83. A magnified glue joint showing glue dowels.

on each side of the joint. The pores filled with glue act as minute dowel pins. The thousands of tiny dowel pins, drying in the wood and being connected to the thin layer of dried glue in the joint itself, fasten the pieces firmly together, as shown in Figure 83.

Kinds of Glues. Casein and synthetic resin glues are the most important for use in aircraft construction and repair. Blood albumin glues were formerly used to a considerable extent for water-resistant plywood, but they have largely been displaced by synthetic resin glues. Blood albumin glue may be used, however, in some cases as a filler or extender for synthetic resin glues. Animal and vegetable glues used in the construction of furniture are not suitable for aircraft work. Most glues used in ordinary fabrication and repair of aircraft are coldpress glues. Marine glues are sometimes used in wooden float and hull construction. Marine glues are used more generally to produce a watertight joint than to make a joint where high strength is required.

Casein Glues. Casein glues, which are prepared largely from the curd of milk, usually contain other material, such as lime or sodium salt, to help dissolve the casein and make a more desirable mixture of the glue. In using casein glues, the manufacturer's directions for mixing and application should be closely followed.

Mixing Casein Glues. For casein glue, a ratio of 1 part of dry glue to 2 parts of water by weight produces a mixture suitable for most joints. Some variation in the proportions may be necessary for particular types of joints in order to get the best results. Any variation from the manufacturer's directions should be made only by a thoroughly experienced workman. In mixing prepared casein glues, the water should be placed in a mixing bowl first, and the glue sprinkled or sifted in slowly while stirring rapidly. Figure 84 shows a simple balance which may be made by the workman to obtain the proper mixture of water and glue. There are many types of mechanical mixers which may be used, although for small amounts an ordinary wooden paddle, operated by hand, may be satisfactory.

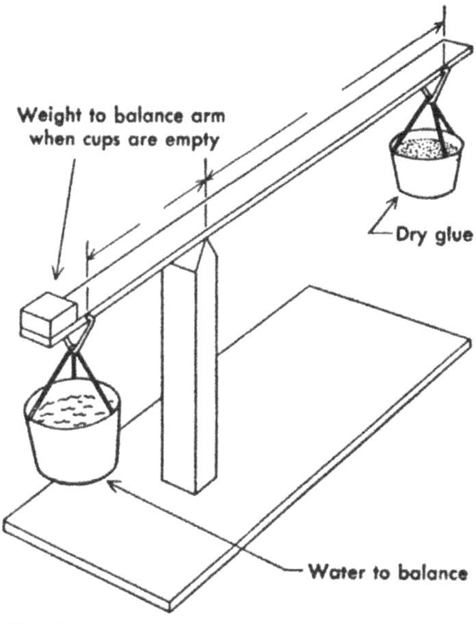

Fig. 84. A homemade balance for weighing water and glue powder.

When a mechanical mixer is used, the mixing paddle should turn at a rate of 100 to 120 revolutions per minute. After all of the glue has been added, the mixer should be slowed down to about one-half this rate and stirring continued until all particles of the glue have been dissolved, and a smooth mixture results. This mixing usually requires from 20 to 30 min. Most casein glues thicken after this amount of stirring and should be allowed to set for 20 to 30 min. and then restirred. After the second stirring, the mixture should be about the thickness of rich cream. If the mixture shows lumps at the end of the mixing it should be discarded. The glue should not be stirred rapidly enough to cause foaming. Casein glue should be mixed at not less than 70°F.

A satisfactory mixer may be improvised by placing a small wooden paddle, as shown in Figure 85, in the chuck of a drill press. This is simply a piece of dowel which has a slot sawed in one end and a thin piece of wood inserted into this slot and fastened. The other end is shaped to fit the drill chuck.

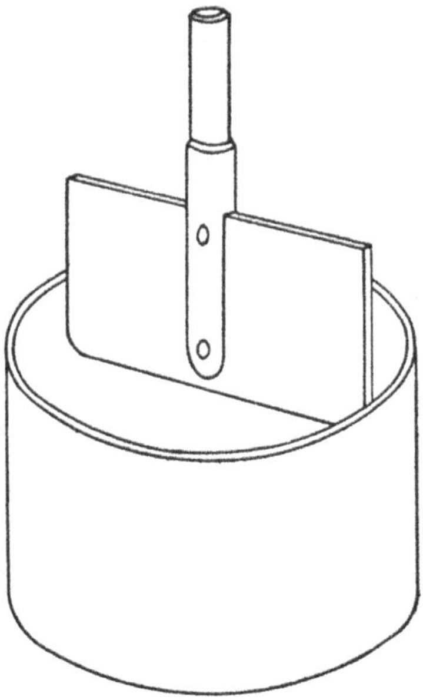

Fig. 85. A homemade drill-chuck paddle for mixing glue.

Casein glue mixed more than 8 hr. in winter or 4 hr. in summer should not be used, as it loses its quality.

The most commonly used implement for applying glue in repair or fabrication of structural members is a stiff brush. A knife or paddle may be used to force the glue into seams or cracks. When glue of the proper consistency is spread over two surfaces to be joined, and pressure is applied, glue enters the wood pores, forming the microscopic dowel joints. The wood pieces are held together by minute glue pins which clamp firmly on either side as they harden.

Proper pressure for glue joints should be strictly followed, and the joint to be glued must be free from any foreign substances which would prevent the glue from entering the pores of the wood.

If the surface cannot be thoroughly cleaned either by planing away damaged wood or foreign material or by gasoline or carbon tetrachloride, the wood should be discarded. Table VI gives the characteristics of glues most commonly used in construction and repair of aircraft.

Cleaning Gluing Equipment. Mixers, spreaders, brushes, glue pots, or any other receptacles which have come in contact with the glue must be thoroughly cleaned at least once each day. Any particles of old glue in freshly prepared mixtures will cause a deterioration of the glue. Paper cups which have been coated with paraffin may be used by individual workmen and discarded at the end of the day or when a new batch of glue is prepared. Copper and brass containers should not be used for mixing glues. Do not use soft paint brushes for glue. Use only stiff, vegetable or synthetic-fiber brushes. These are more efficient, allow better control of spread, and are easier to clean. Wash all brushes every one or two hours (before the glue jells), even though work is incomplete. When washed often and regularly, brushes are more efficient, easier to clean, and last longer. Use plenty of water or alkaline washing solution and remove every trace of glue. The water may be any temperature. If done quickly, boiling water may be used — it will liquefy and wash away resin glue before it can set.

Wash all glue from clothing or skin before the glue dries, if possible. This may require frequent hand washing, but this practice promotes

TABLE VI. CHARACTERISTICS OF GLUES MOST COMMONLY USED IN CONSTRUCTION AND REPAIR OF AIRCRAFT

PROPERTY OR CHARACTERISTIC	CASEIN GLUE	BLOOD ALBUMIN GLUE	SYNTHETIC RESIN GLUE
Strength (dry)	Very high to high	High to low	Very high to high
Strength (wet, after soaking in water 48 hr.)	About 25% to 50% of dry strength; varies with glue	About 50% to nearly 100% of dry strength	Very high; nearly 100% of dry strength
Durability in 100% relative humidity or prolonged soaking in water	Deteriorates eventually; rate varies with glue	Deteriorates slowly, but completely, in time	Very high if resin is unadulterated
Rate of setting	Rapid	Very fast with heat	Very fast with heat
Working life	Few hours to a day	Few to many hours	Few to several hours for liquid forms; several weeks for films
Consistency of mixed glue	Medium to thick; little change with temperature	Thin to thick; little change with temperature	Medium for liquid forms
Temperature requirements	Unimportant	Heat required to set most glues	Heat required for most glues
Mixing and application	Mixed cold with water; applied cold by hand or mechanical spreaders	Usually mixed cold with water; applied cold by hand or mechanical spreaders	Applied as received or after addition of catalyst; liquid forms best when applied by rubber-covered rolls
Tendency to foam	Slight if not mixed too rapidly	Slight to pronounced	Slight
Tendency to stain wood	Pronounced with certain woods	None, except dark glue may show through thin veneers	None, although glue may penetrate through thin or porous veneers
Dulling effect on tools	Moderate to pronounced	Slight	Moderate
Spreading capacity:			
Extremes reported	35 to 80 sq. ft. per lb.	30 to 100 sq. ft. per lb.	30 to 100 sq. ft. per lb.
Common range	40 to 60 sq. ft. per lb.	———	35 to 50 sq. ft. per lb.

greater efficiency and helps prevent injury to the skin. If allowed to dry on the skin, the glue is difficult to remove. Casein glue will soften on soaking, but urea or phenol resin must be peeled or worn off. The use of a protective salve often prevents glue from sticking to hands.

To remove dried urea resin glue from fabrics, boil the glue-soiled portion of the fabric until the glue is soft enough to be brushed out. It is usually impossible to remove dried phenol resin glues from fabrics. Any chemicals which would soften the glue will generally destroy the fabric. Care should be taken to avoid having the dry powdered glues, particularly the urea or phenol glues, come into contact with skin, eyes, or mucous membranes of the nose or throat.

Preparation of Wood for Gluing. The moisture content of the wood to be glued is important. The moisture content should be from 8 to 12 per cent for stock ½ in. thick or thicker. A moisture content of 5 to 10 per cent is satisfactory for veneers or thin laminations. Wood to be glued should be free from case hardening and other stresses. The wood should be surfaced smoothly, free from scratches or machine markings and any dust or fine particles of wood such as are left by sandpaper.

Gluing Conditions. To produce strong joints, a sufficient amount of glue must be spread on the wood surface. The glue should be of proper consistency, and the proper pressure applied. "Starved" or "crazed" joints may occur where not enough glue is applied. A "dried" joint may occur when the glue is too thick and too much glue is applied for the amount of pressure. A good joint should have practically no free glue in the joint. A well-glued joint consists of a thin layer of glue, most of which is forced into the wood by pressure applied.

Amount of Glue Required. For casein glue, about 7½ lb. of wet glue will cover approximately 100 sq. ft. of joint area, or approximately 1 oz. of wet glue per square foot. This glue should be spread evenly on one of the two contact faces of the joint. Sometimes both contact faces are coated, but only about 25 per cent more glue should be used than would be spread on one face. A small excess of glue is safer than putting on too little. Under normal conditions, a small amount of glue should be squeezed out of the side of the joint. This glue should not be wiped off, but allowed to dry. The glue entering the wood may draw some of this glue into the edge of the joint, preventing air pockets. If enough glue is squeezed out to run down over the part being glued, the excess may be removed. A small excess of glue along the edge of the glued surface may prevent starved joints.

PROPERLY GLUED JOINT

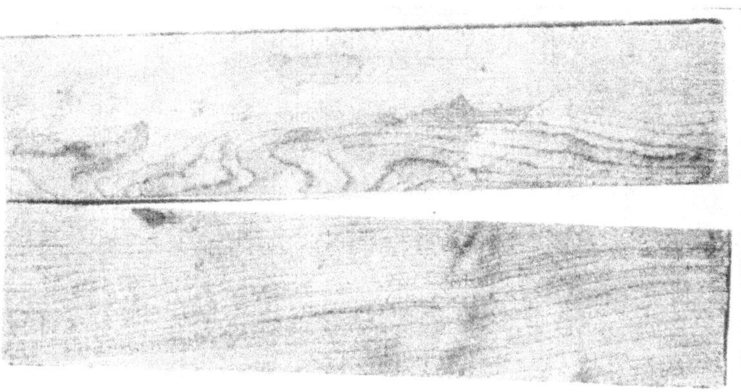

STARVED OR CRAZED JOINT

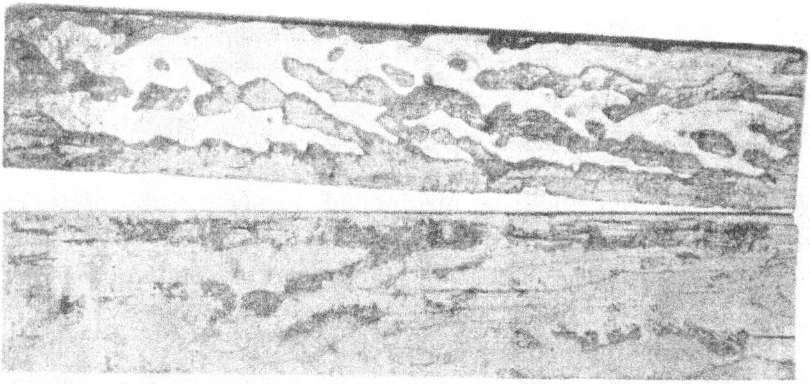

DRIED JOINT WITH THE GLUE TOO THICK

Fig. 86. Glued joints. (Courtesy Forest Products Laboratory)

Assembly Time. The more rapidly the joint is closed and put under pressure after the glue is applied, the better. For most work, 5 or 10 min. is all the time which should be allowed to elapse between the spreading of the glue and the application of pressure. Glued joints, while setting, should not be exposed to changes in temperature. A temperature

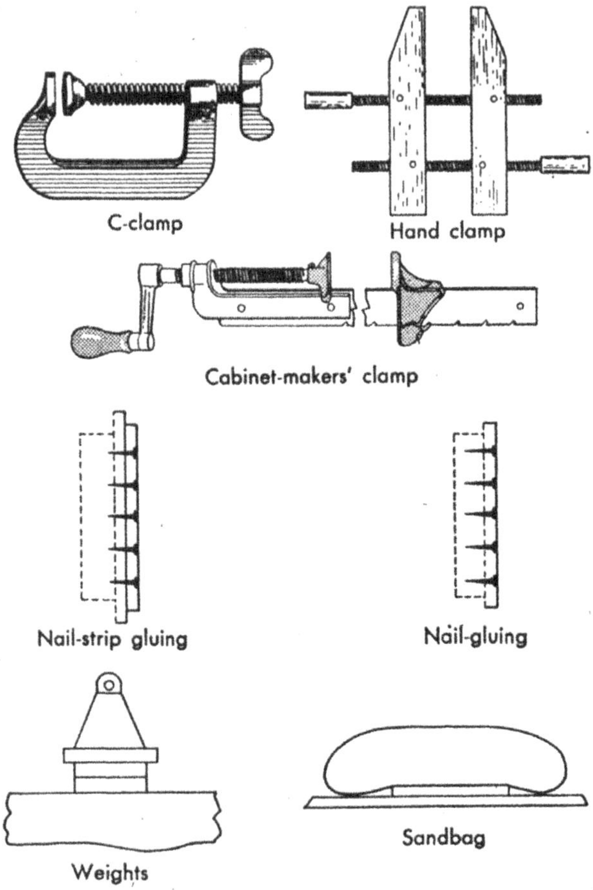

Fig. 87. Methods of applying pressure to glued joints.

of 70°F., or above, should be maintained throughout the preparation of the joint and during the setting time.

Pressure and Clamping. There are many ways of applying pressure to glued joints, but the use of hand clamps, cabinet-makers' clamps, and C-clamps are the most common devices.

In applying pressure, backing boards on each side of the joint should be used. These boards should be of hard, stiff wood such as oak. The backing board should be equal in thickness to the material being glued.

These boards should be heavy enough so that there is no bending or distortion due to the pressure of the clamps. The backing board should extend a sufficient distance on each side of the joint to prevent any slipping of the joint.

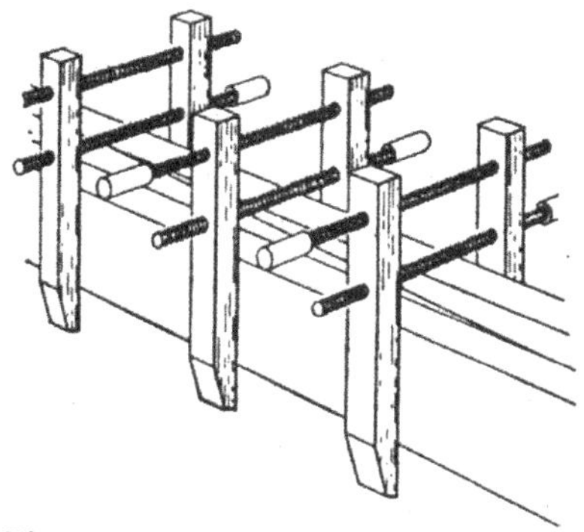

Fig. 88. Solid backing boards prevent slipping of glued joints.

Where large, laminated structures, such as spars, are glued in factory production, special clamp jigs may be used. The amount of pressure applied varies with the kind of wood. For softwood, such as spruce, a pressure of 100 to 150 lb. per sq. in. should be used. For hardwood, such as ash or oak, a pressure of 150 to 200 lb. per sq. in. should be used. For softwoods, such as balsa and basswood, 50 lb. per sq. in. may be sufficient.

Measuring Pressure Applied. Experience is the best teacher of the application of proper pressures. The beginner can experiment by placing a board across a platform scale and, by the use of clamps, figure rather closely the amount of pressure he can apply with a single clamp, remembering that the total pressure is spread over the entire bearing surface. For example, if a small C-clamp exerts 1000 lb. pressure when tightened by hand and this pressure is applied to an area of 10 sq. in., the pressure per square inch would be 100 lb. The general tendency is to apply too little pressure rather than too much. The pressure should be evenly applied to the surfaces being glued, and clamps should be tightened as

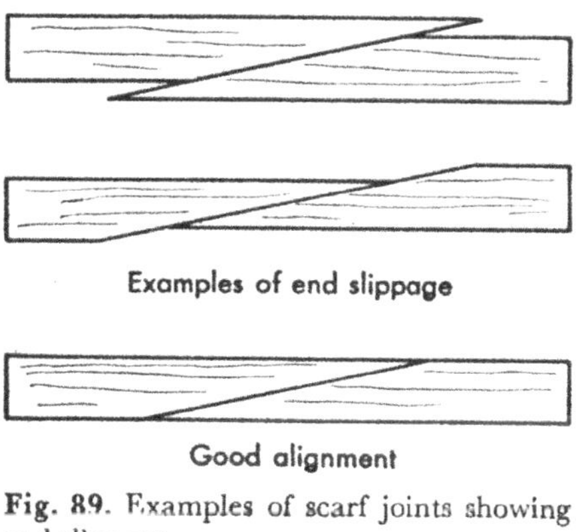

Fig. 89. Examples of scarf joints showing end slippage.

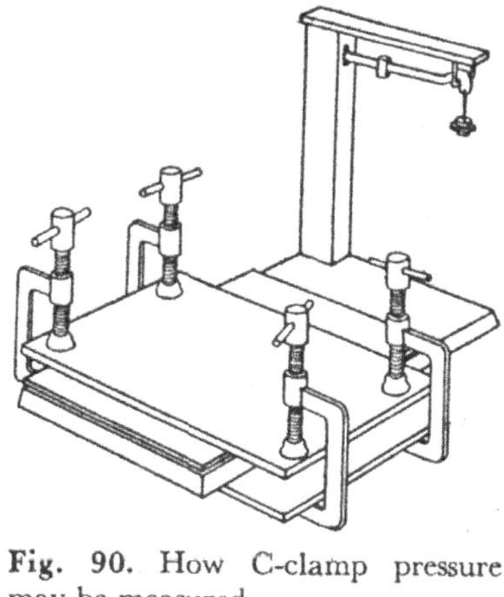

Fig. 90. How C-clamp pressure may be measured.

the glue squeezes out from between the layers, distributes itself within the joint, or is pressed into the wood itself. Gluing clamps should be retightened after pressure has been applied — approximately 10 min. on hardwood and 20 min. on softwood.

Time Pressure Is Applied. Casein glue should usually be under pressure for not less than 6 or 8 hr., but it is much better to leave the glue joint under pressure from one day to the next. Very soft wood, when glued, may be relieved from pressure in 3 or 4 hr., but should be allowed to season at least 24 hr. before being worked.

Gluing Different Woods. Where the grain of the wood is dense, a smaller amount of glue may be used if the glued surfaces are smooth and fit closely together. Very soft or open-grained wood requires larger quantities of glue. Before gluing any kind of wood with which a workman is not entirely familiar, it would be well to glue two or three sample joints under varying conditions to determine the most desirable method.

For softwoods, glue of a medium thickness with a normal spread is recommended. The pressure applied should be from 100 to 150 lb. per sq. in. In gluing, the following woods are usually classified in the softwood group: basswood, cottonwood, noble and white fir, western hemlock, ponderosa, sugar and white pine, Port Orford white cedar, western red cedar, redwood, red, Sitka, and white spruce, and yellow poplar.

The hardwood group requires a glue somewhat thicker with a normal spread, but the pressure should be from 150 to 250 lb. per sq. in. The following woods are usually classified in the hardwood group: white ash, beech, blackgum, yellow birch, black cherry, Douglas fir, American elm, rock elm, hickory, mahogany, magnolia, soft maple, hard maple, red oak, white oak, sweetgum, sycamore, black walnut, and water tupelo.

Scarf Joints and End-Grain Gluing. Particular care must be taken on all joints where end-grain wood is glued. All scarf joints are cut at an angle to the grain, and, since this type of joint is most often used in aircraft construction, great care must be taken when gluing it. Usually, joints of this kind should be sized with a glue mixture somewhat thinner than that used for regular gluing. A mixture of 1 part of glue to 3 parts of water has been found to be satisfactory. A sizing coat is a coat of glue which is applied and allowed partially to dry before the joint is glued in the regular manner. The mixture for the final gluing should be somewhat heavier than for ordinary gluing. About 1 part of glue to 1⅛ parts of water by weight should be used for casein glue. Both surfaces of the joint should be covered with this glue and a pressure of approximately 200 lb. per sq. in. applied. This method should be used wherever end-grain gluing is necessary. Glued joints should be allowed to dry thoroughly, so that any moisture added to the wood during the gluing process is evaporated before the final finishing or machining of the joint.

Time for Conditioning after Gluing. For propellers and similar heavy construction where the amount of wood is large in proportion to the amount of glue, it is not necessary that all of the moisture be eliminated. For laminations of ¾ in. or more, a conditioning period of 7 to 10 days at room temperature is recommended. For quick-drying woods, such as spruce or pine, 2 to 3 days should ordinarily be sufficient for conditioning. Where thin layers of wood are glued together, it is often advisable to dry out the surplus moisture by artificial methods, either by blowing warm air over the parts or placing in a room having a comparatively high temperature. Temperatures of 100° to 110°F. ordinarily cause rapid drying.

Synthetic Resin Glues. Synthetic resin glues are based on resins made artificially from various chemical components. Phenolresin (warm setting) and urea resin (cold setting) are the synthetic glues most commonly used in woodwork. In the preparation of these glues, the reaction between the chemicals used in their manufacture is stopped before it is complete. The glue is applied to the parts to be glued, and the reaction is completed during the gluing operation. Some glues require the addition of a chemical substance to cause hardening. These substances cause the completion of the reaction. Most plywoods are at present made with the synthetic resin glues by the hot-press method.

Mixing Synthetic Resin Glues. The manufacturer's directions for mixing these glues should be followed carefully. The quantities of the different materials should be weighed carefully, as some parts may be added in comparatively small quantities. The powdered glues are mixed with water and hardeners in accordance with the manufacturer's instructions. Conditions of temperature and pressure should be those recommended by the manufacturer.

XI JOINTS AND LAMINATIONS

The most important joint used in aircraft construction and repair is the scarf joint. The scarf joint is made by beveling off the ends or sides of the pieces to be joined together at a definite slope. This slope may be as shallow as 1 in 20, and is rarely steeper than 1 in 10. Scarf joints usually entail gluing end grains together. Since gluing end grains of wood together is not generally satisfactory, all scarf joints must be so

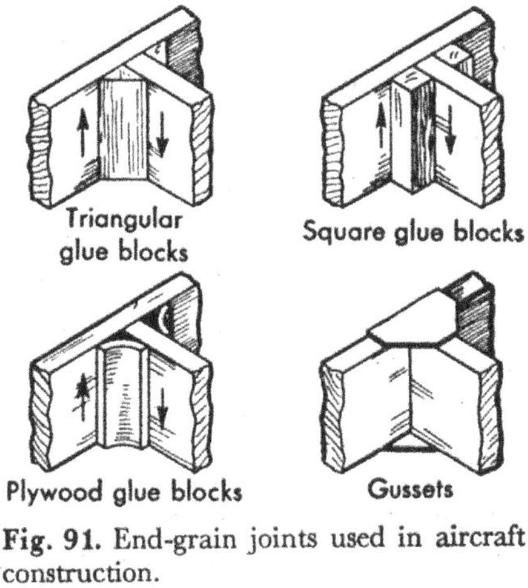

Fig. 91. End-grain joints used in aircraft construction.

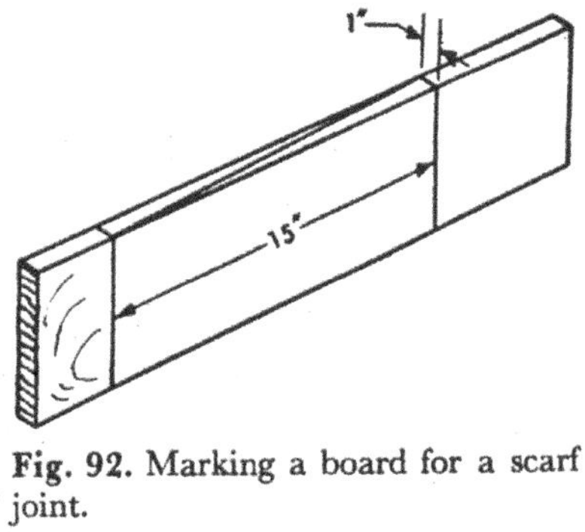

Fig. 92. Marking a board for a scarf joint.

cut that the slope of the joint is great enough to overcome the disadvantage of end-grain gluing. Scarf joints are used in splicing all wood parts in the aircraft structure. The woodworker who works on aircraft should be expert in the preparation of scarf joints.

It is sometimes necessary to join the end of a piece of wood to the side of another piece. This is usually done by the aid of glue blocks of the types shown in Figure 91. In wing-rib structures, the joints are strengthened by plywood gussets. Nails are often used in aircraft structures but are not depended upon to add strength to the structure. They are used to hold parts together until the glue has set. Nails and screws in aircraft structures tend, because of moisture changes and vibration, to loosen and pull out when placed at points where stresses occur. End-grain nailing is unsatisfactory where there is any tensile stress.

Most wood joined in aircraft construction is done with the aid of gluing. Such joints as mortise, tenon, dovetail, and other cabinetmaker joints are not used in aircraft structures due to the difficulty of obtaining

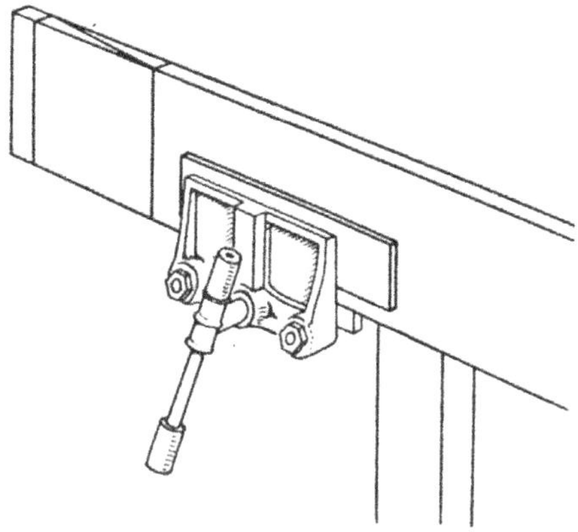

Fig. 93. A board clamped in a vise for sawing a scarf joint.

Fig. 94. Sawing a scarf joint.

proper gluing pressure and tightly fitted joints. These joints tend to loosen when subjected to vibration or torsional stresses.

Making a Scarf Joint. A scarf joint may be made as follows: Select a board 1 × 6 × 30 in. and 4 in. from one end draw a line at right angles to one edge. Since the scarf joint is going to have a slope of 1 in 15, draw another line 15 in. from the first line on the same face of the board.

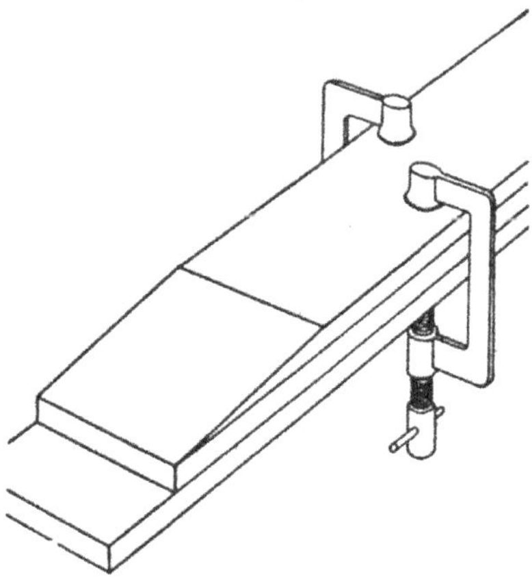

Fig. 95. One-half of a scarf joint mounted on a board for smoothing.

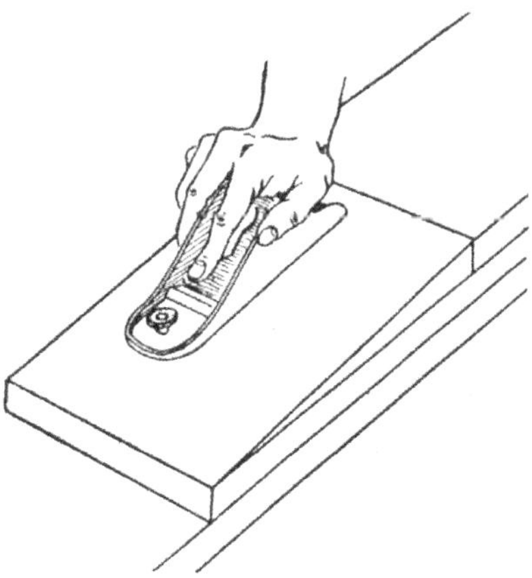

Fig. 96. Use a block plane to smooth a scarf joint.

Draw lines across the edge of the board and join the lines on the edge with diagonals, as shown in Figure 92. It is well to extend the lines completely around the board and draw a corresponding diagonal on the opposite edge. The joint should be cut with a fine-tooth saw. Each side of the joint must be finished smoothly. For this a block plane is most satisfactory. Check the feather edge for squareness with a try square.

Fig. 97. Test the feather edge for squareness with a try square.

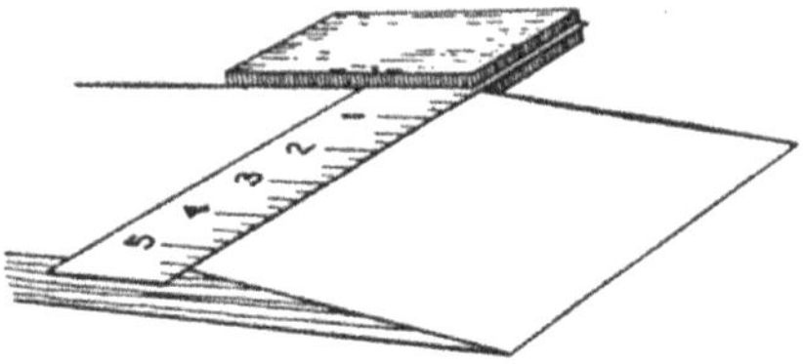

Fig. 98. Test the heel of a joint for squareness with a try square.

Check the heel for squareness in the same manner. If the feather edge is not square, additional material must be removed. This may make it necessary to remove some material back from the heel as the slope must be 15 in. long when smooth and ready for gluing. In order to support the feather edge, the board may be mounted on another board as shown in Figure 95. Test the joint surface for flatness by a steel straightedge. The joint should be flat both lengthwise and crosswise. Prepare the other half of the joint in the same manner. If the slope is not exactly

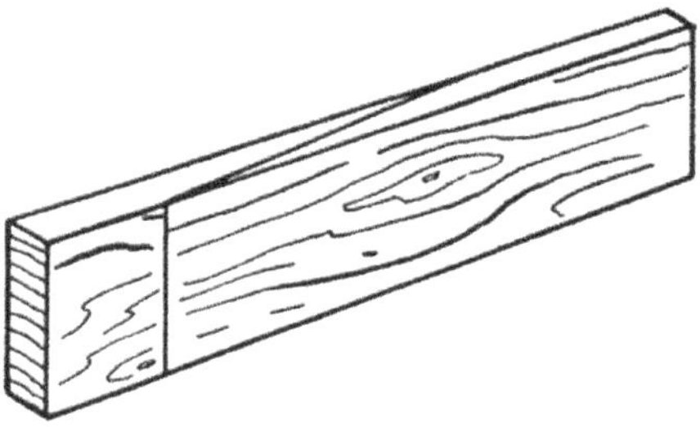

Fig. 99. A finished scarf joint.

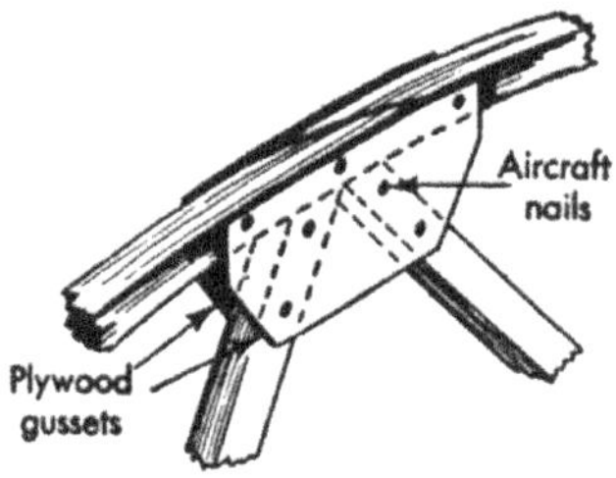

Fig. 100. A joint reinforced with nail-glued gussets.

true on each piece, the glued joint must be trued or the piece will not be straight after gluing. Figure 89 (page 91) shows faulty and good joints.

Care must be taken in assembling the joints so that there is no end slippage after the glue has been spread on both surfaces of the joint. The proper glue blocks and clamps are applied to the joint as shown in Figure 88. In gluing a joint of this kind, a piece of paper is placed between the material being glued and the glue blocks to prevent the glue sticking to the blocks when dry. Leave the joint in the clamps preferably overnight but not less than 8 hr. Remove the clamps and, with a chisel and hammer, try to separate the joint to determine whether or not a properly glued joint has been made.

Laminations. Laminated woods differ from plywoods in that the grain in the various laminations generally runs in the same direction in each piece. In plywood the alternate layers have the grain at right angles. The laminations in laminated members are usually much thicker than the veneers used in plywood.

Such parts as wing-tip bows are built up in the form of laminated material made from several thin layers of wood. Laminated structures which are curved are glued in the form of the curve, the glue binding the wood to the proper shape. These shapes should be left in the gluing form until thoroughly dried. Wood parts made by laminating include fuselage-forming members, bulkheads, floats, hulls, and spars.

XII AIRCRAFT WOODWORKERS' TOOLS

It would be impossible to carry on our civilization without the assistance of the proper tools. Many of the more complicated machines and structures are built by means of simple tools properly handled by a skilled workman. An old saying is that "a workman is known by his tools."

The poor workman will give his tools poor care. The good workman will give his tools the finest care and will always be found using the proper tool to do the job.

Tools should always be cleaned, sharpened, oiled, and used properly. There is a right and a wrong way to use every tool. It is of great importance for the beginner to learn to use and care for each tool in a proper manner.

The following figures show the most important tools used by the aircraft woodworker.

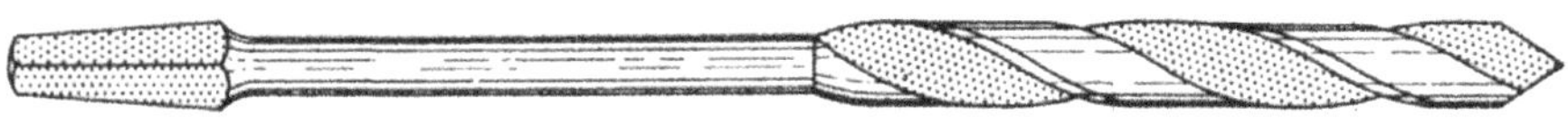

Fig. 101. A drill for airplane woodwork.

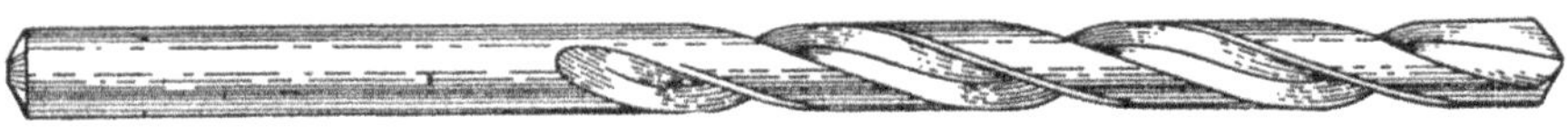

Fig. 102. A bitstock drill.

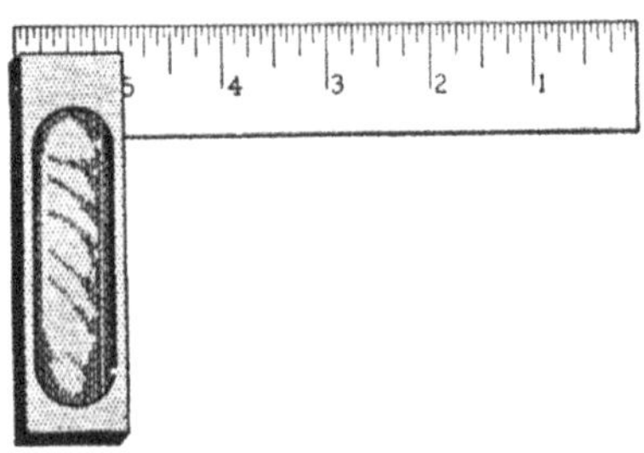

Fig. 103. A try square.

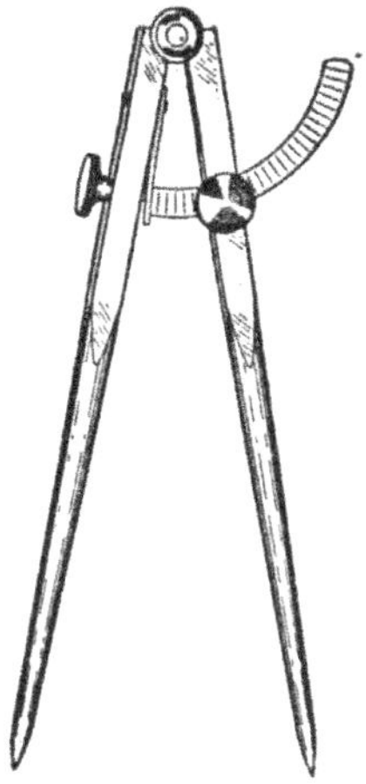

Fig. 104. A wing divider.

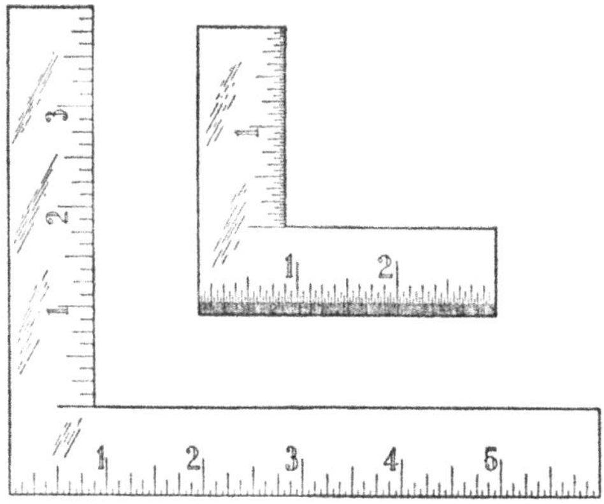

Fig. 105. Steel squares.

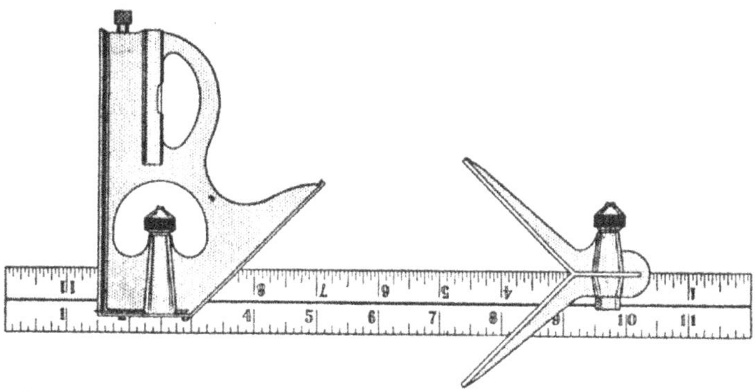

Fig. 106. A combination square.

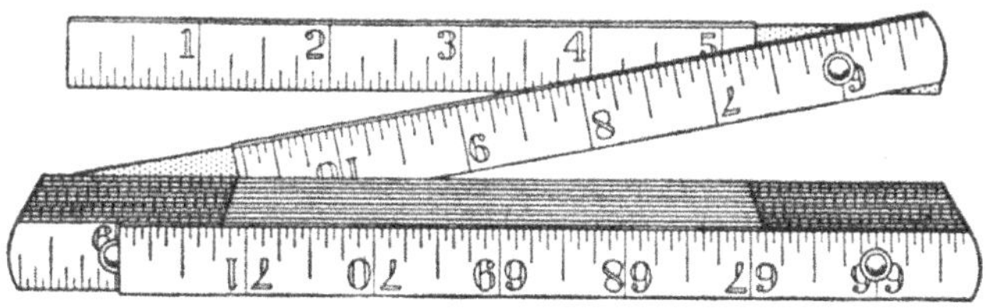

Fig. 107. A folding rule.

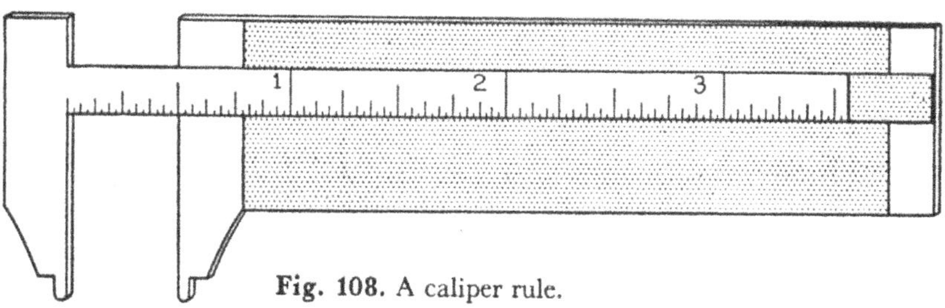

Fig. 108. A caliper rule.

Fig. 109. A pin punch.

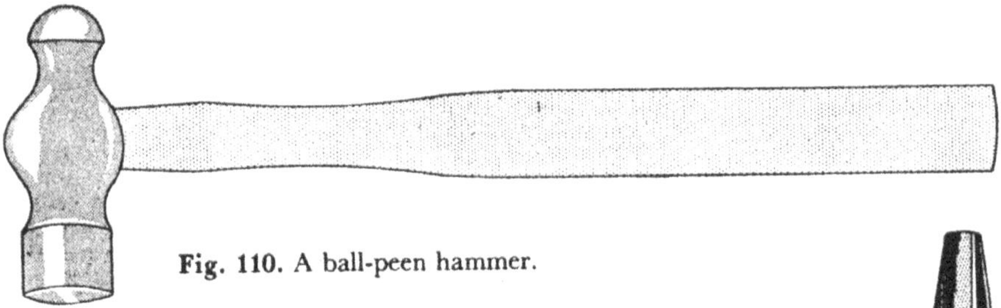

Fig. 110. A ball-peen hammer.

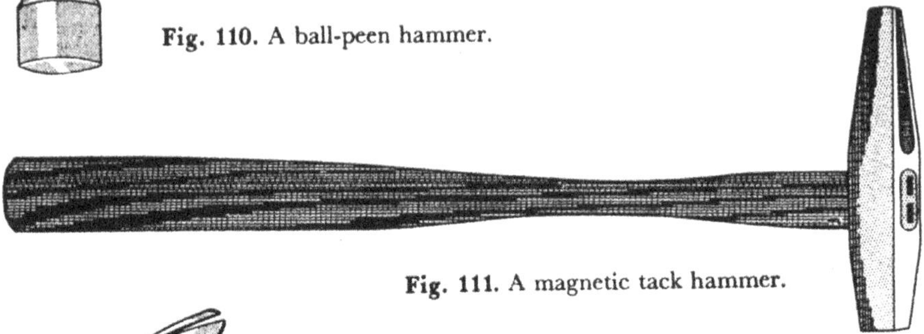

Fig. 111. A magnetic tack hammer.

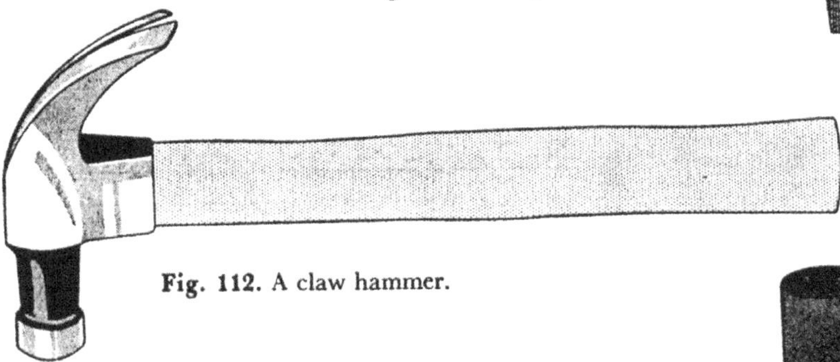

Fig. 112. A claw hammer.

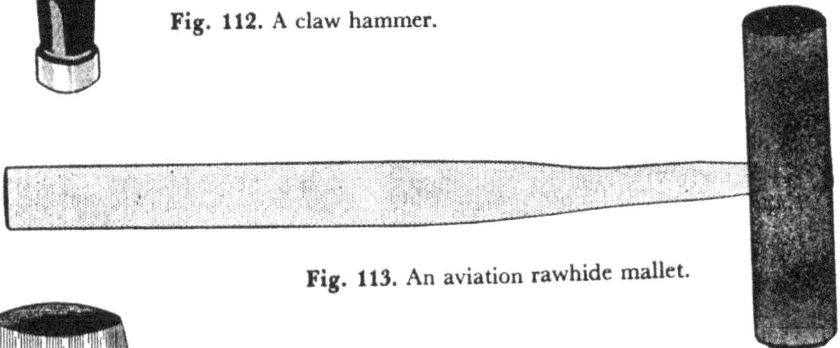

Fig. 113. An aviation rawhide mallet.

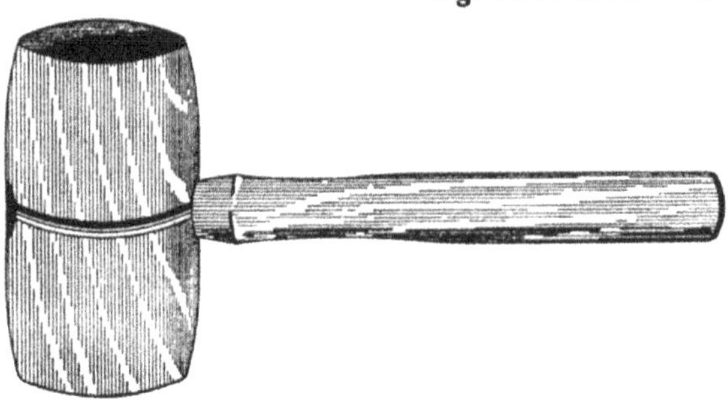

Fig. 114. A wooden mallet.

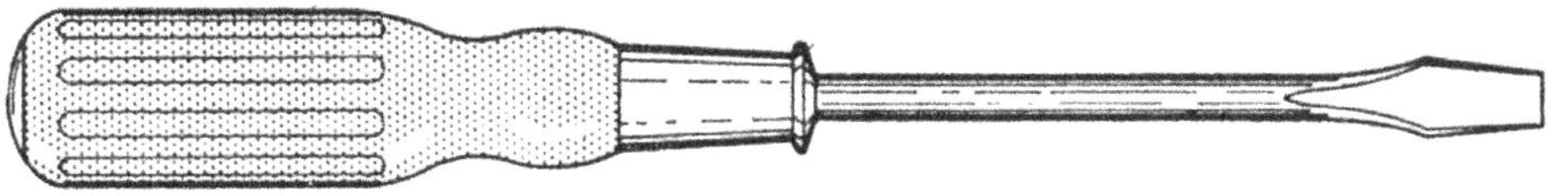

Fig. 115. A screw driver.

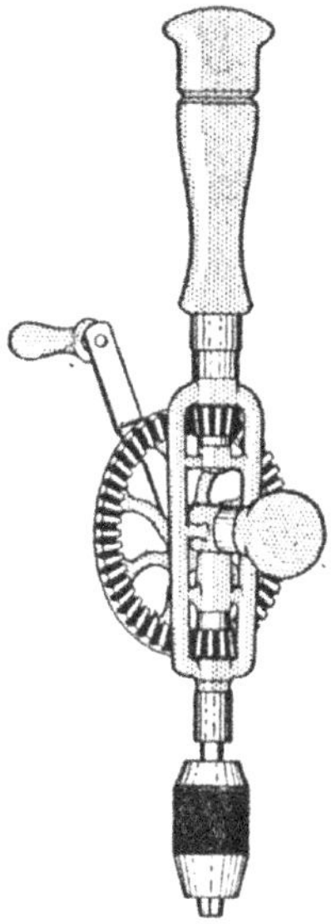

Fig. 116. A hand drill.

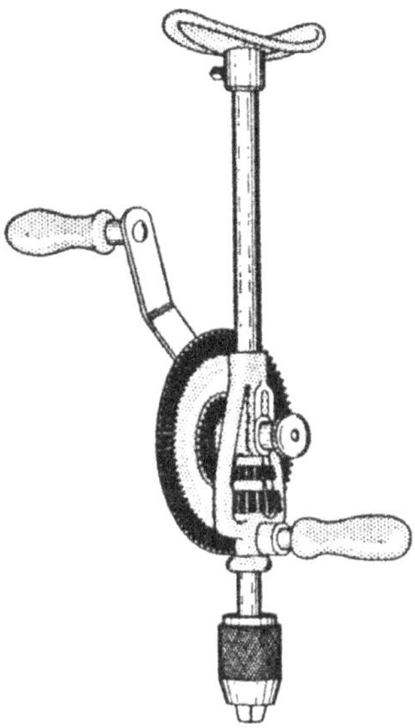

Fig. 117. A breast drill.

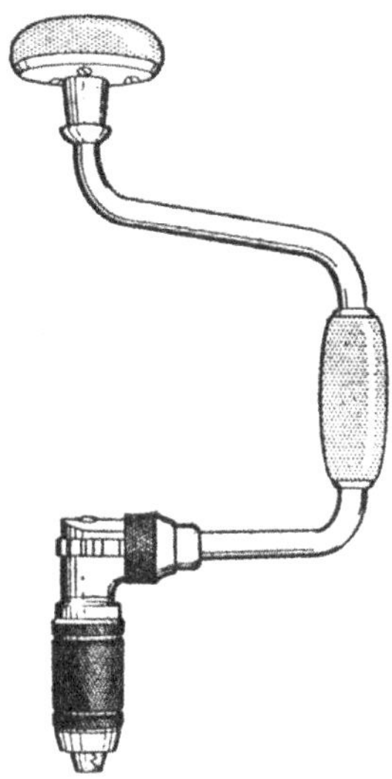

Fig. 118. A bit brace.

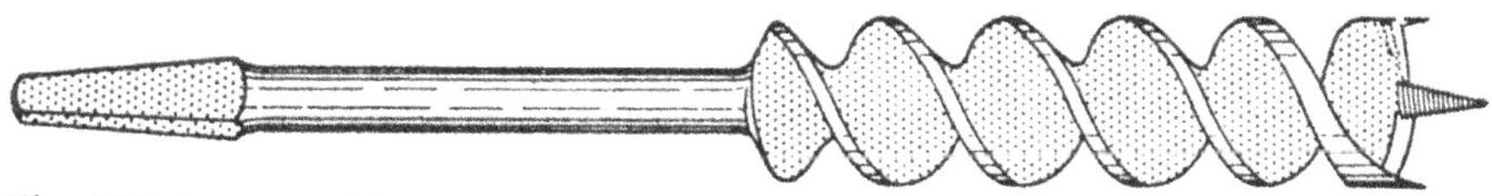

Fig. 119. An auger bit.

Fig. 120. A wood drill for a brace.

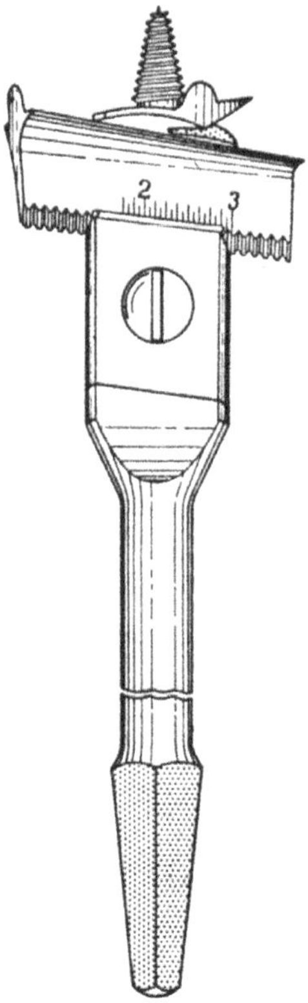

Fig. 121. An expansion bit.

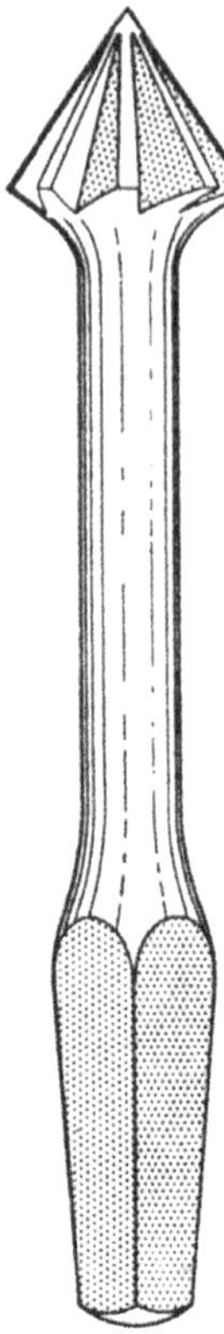

Fig. 122. A rose countersink.

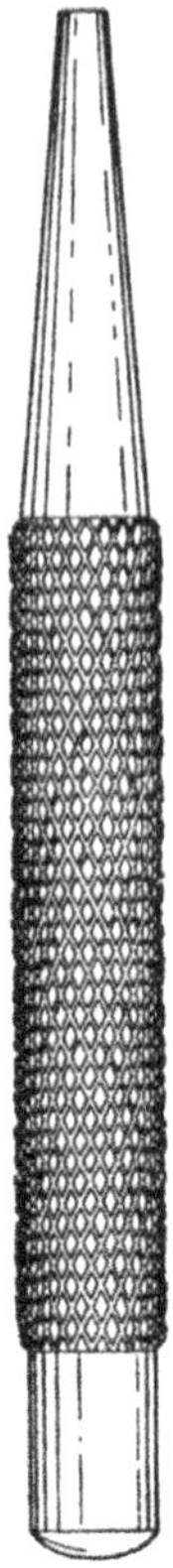

Fig. 123. A nail set.

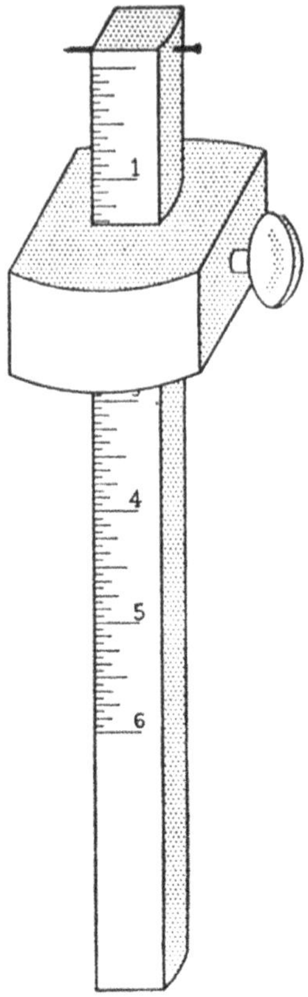

Fig. 124. A marking gauge.

Fig. 125. A spokeshave.

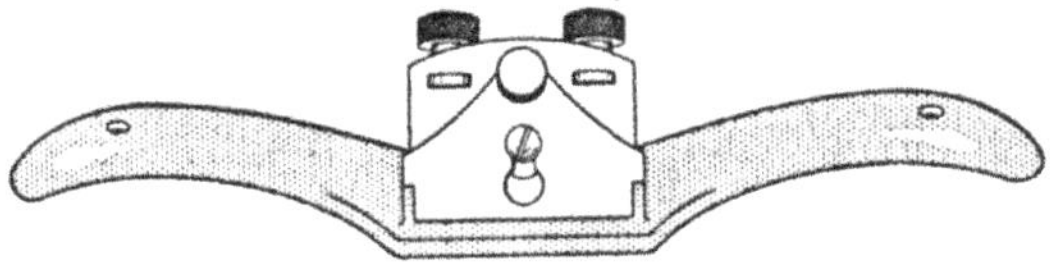

Fig. 126. A draw knife.

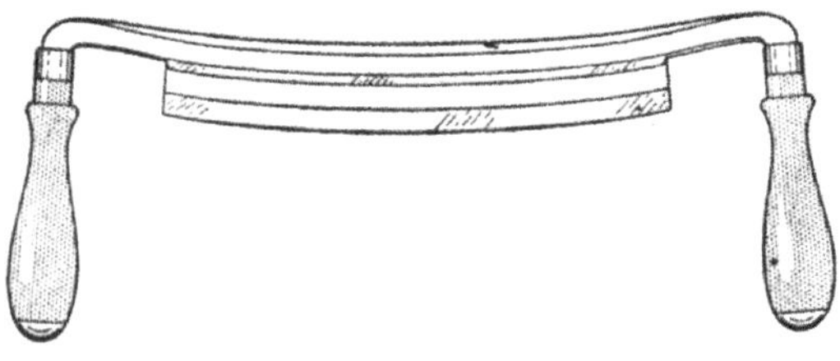

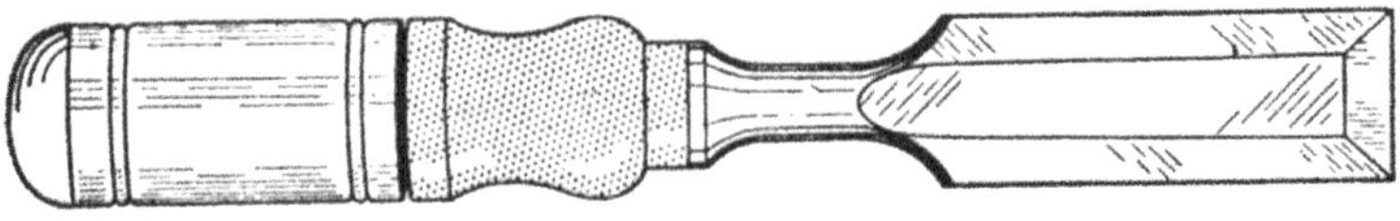

Fig. 127. A Stanley paring chisel.

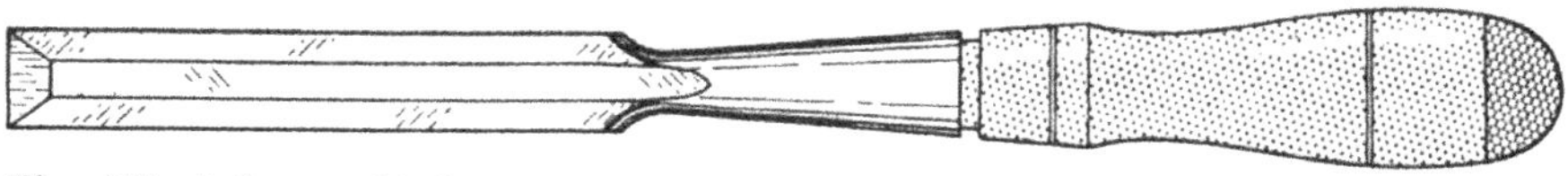

Fig. 128. A firmer chisel.

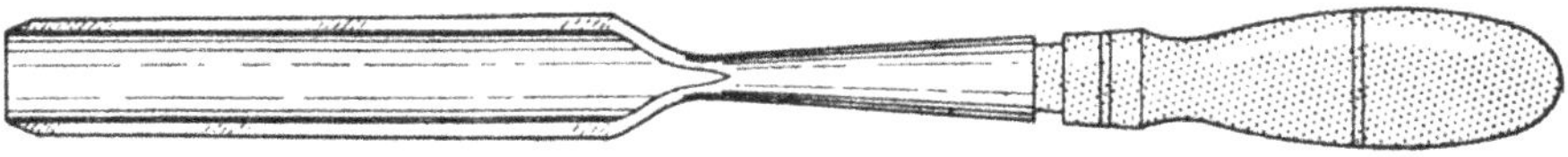

Fig. 129. A firmer gouge.

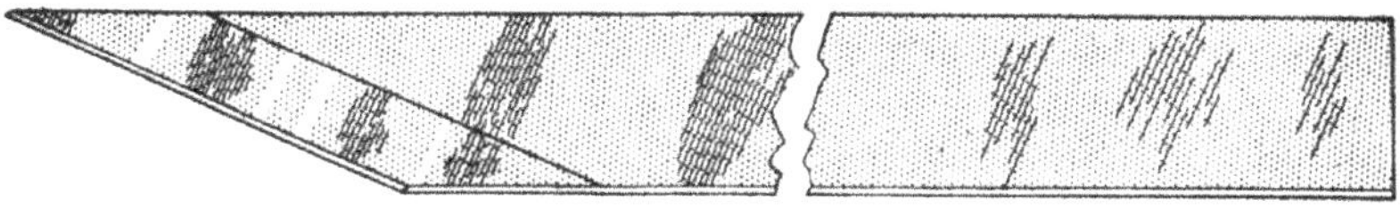

Fig. 130. A bevel knife blade.

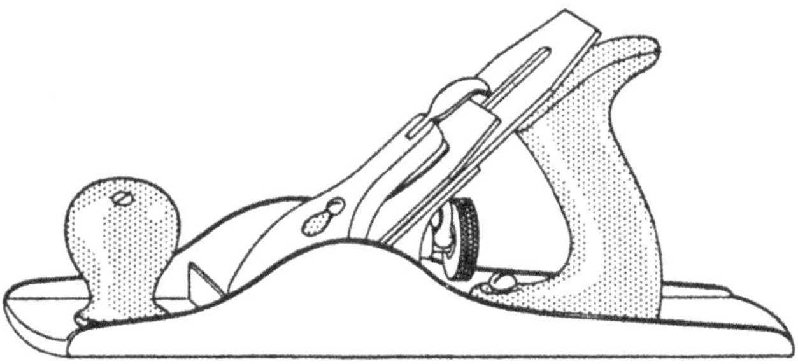

Fig. 131. A jack plane.

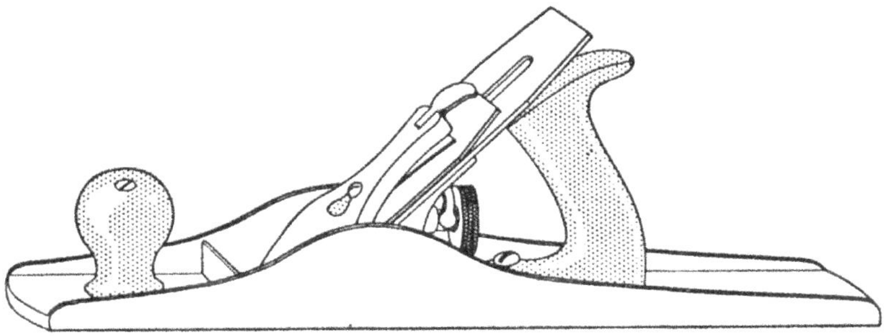

Fig. 132. A fore plane.

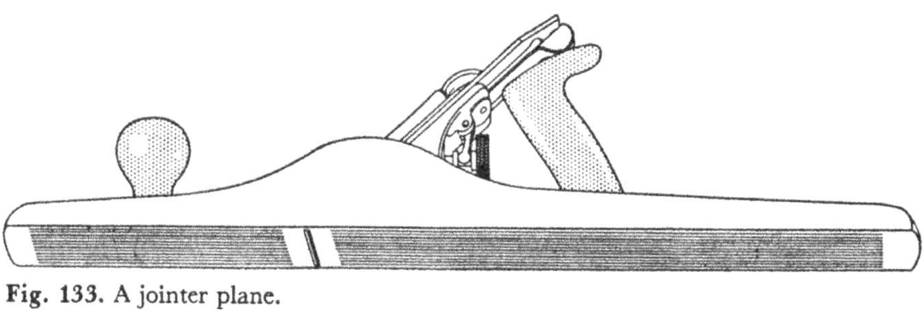

Fig. 133. A jointer plane.

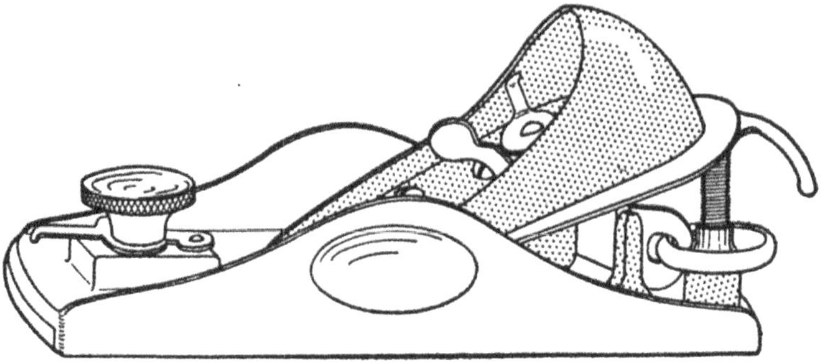

Fig. 134. A block plane.

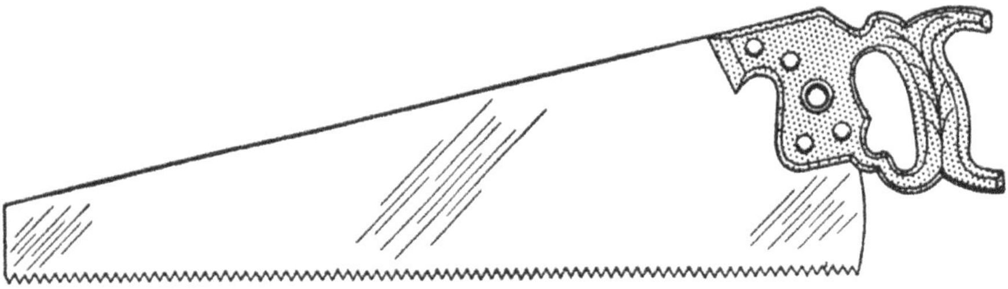

Fig. 135. A handsaw.

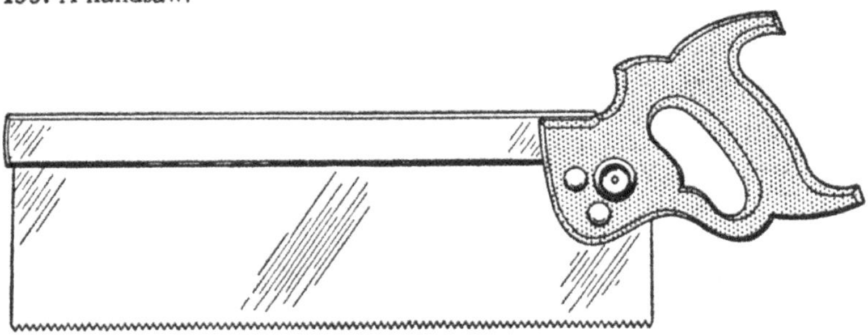

Fig. 136. A backsaw.

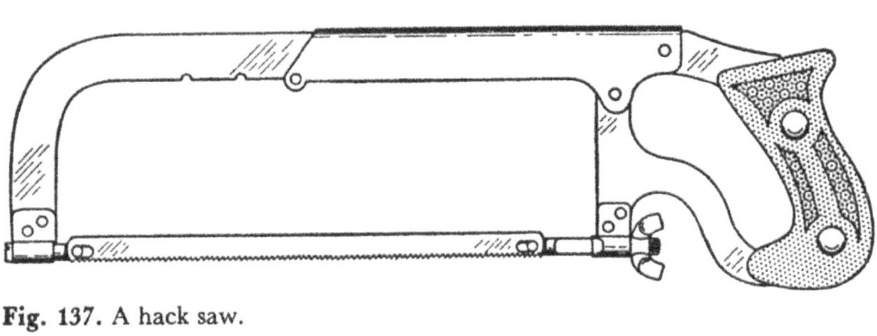

Fig. 137. A hack saw.

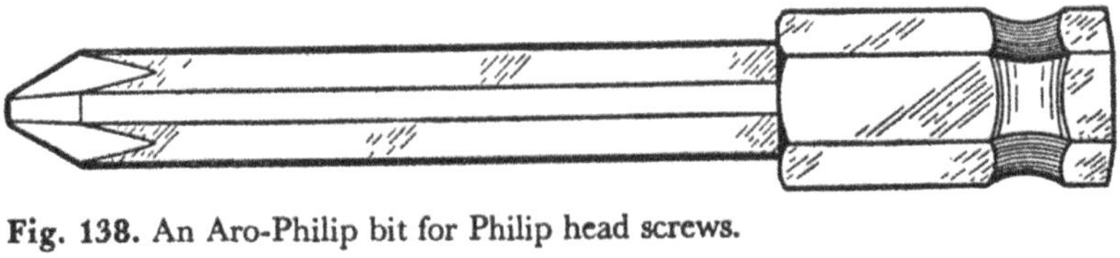

Fig. 138. An Aro-Philip bit for Philip head screws.

XIII AIRCRAFT DOPING AND FINISHING

The technique of doping and finishing an aircraft requires greater care and closer attention than almost any other phase of the manufacturing process. All parts of an airplane must be protected from corrosion, changes in temperature, moisture, etc.

In the manufacturing of airplanes, definite specifications are usually furnished for the types of finishes and protective coatings to be given to each part. These finishes are designed to last during the life of the structure. The finishes must be such that they will stand up against wide changes of climate. The airplane may, in a few minutes' time, be taken from a hot, damp climate such as the tropics, to a subzero, arctic climate at 25,000 or 30,000 ft. of altitude. The finish must be able to withstand not only extremely humid climates, but also the heat, dryness, and sand of desert regions.

The outside finish of the airplane not only must protect the airplane from the elements but must present a pleasing appearance. Interior parts, such as the inside of the wings, control surfaces, and fuselage, must be protected from moisture which may condense within the parts, as well as the loss of excessive moisture when used in hot, arid regions. Considerable trouble has been had with various wooden parts of airplanes in desert regions because of cracking and splitting in the extreme dryness.

Most of the interior finishing is done before the parts are assembled. These finishes should be fast drying and weather and moisture resistant. Spar varnish is commonly used for the finishing of all woodwork. This varnish usually has a resin base, is slightly yellow in color, and transparent enough so that inspection of the wood which it covers may be made without removing the varnish. A paint would cover up or conceal cracks and defects which might appear in service but which can be easily detected through varnishes. It is not necessary to highly finish concealed wooden parts, as protection from the elements is more important than a pleasing appearance.

Any surface which is to be glued should not be covered with varnish or any other finish. These surfaces can be masked off with masking tape or any suitable material while the part is being varnished.

Fig. 139. Dipping an entire wood wing into the finishing solution. Note the construction of the wing. (Courtesy Forest Products Laboratory)

Varnish dries by a chemical action on exposure to the air. Varnish should be applied in thin coats to allow proper drying. When a thick coat of varnish is applied, the outside of this coat may dry thoroughly, preventing the oxygen in the air from contacting the varnish underneath. This would cause peeling or rubbing off in service. Varnish is usually used as furnished by the manufacturer, although often, for the first coat, it is thinned with turpentine.

It is necessary to protect metal parts which come in contact with wood from the moisture in the wood. This is done by the application of a suitable sealing material which does not harden completely. Metal parts are covered with special metal sealers or primers to protect them from corrosion.

Dope used for finishing fabric should never be placed over varnished surfaces as it causes varnish to soften and peel. Dopeproof paint is usually pigmented white and has a resin base and is thinned with water instead of the usual paint thinner. For the final finish on exterior surfaces, a pigmented dope or varnish is used. Aluminum powder and lead chromate are two common pigments.

Aircraft Skin. The outer covering of an airplane is usually called the "skin." The skin is a tight, weatherproof covering which protects the inner parts of the airplane and gives shape to the structure. When fabric is used it accomplishes these purposes. If plywood or other kinds of stiff covering are used, they may also add strength to the airplane structure. When the skin adds strength to the structure it is called "stressed skin." The skin may be made from fabric, plasticized plywood, plywood, aluminum alloy sheets, or sheets made from various alloys of steel.

The fabric used on most airplanes is of long-staple cotton which has been closely woven and is free from fuzz. When cotton fabric is treated with dope, it shrinks and becomes waterproof and resistant to decay.

Acetone is usually used as a dope solvent, and ether is sometimes added as a thinner. If it is necessary that the doped fabric be somewhat flexible, a small amount of castor oil can be added to the mixture. Two standard dopes are nitrate and acetate cellulose compounds.

The plywood covering of aircraft structures is very often covered with fabric to increase its strength and protect the plywood from the effects of weather.

Wood which is to be covered with doped fabric is usually given a coat of dopeproof material.

Fabric Finishes. There are two general types of finish used on fabrics: (1) a pigmented dope finish, and (2) a semipigmented dope finish.

Pigmented Dope Finish on New Fabric. The following method, if carefully followed, will produce a durable and highly satisfactory pigmented dope finish. After the fabric has been placed on the structure in accordance with approved methods, a coat of clear dope is applied. The dope is applied with a 3- or 4-in. wall brush, and a full wet brush of dope should be used. The dope should be applied evenly over the surface being doped. Do not try to brush the dope thin or to cover too much space with one brushful of dope. The first coat will anchor the following coats to the fabric. Remember, this is not painting but filling the fabric with dope. Saturate the fabric thoroughly with this first coat and allow to dry from 30 to 45 min., depending on the temperature.

At this point, reinforcing tape, patches, drain grommets, etc., may be applied, either with the second coat or as a separate operation immediately before applying the second coat of clear dope. They are usually applied with the second coat. This is accomplished by doping the area and placing the tape, grommets, etc., on to the wet dope, pressing firmly into place. Be sure no air bubbles are trapped beneath the fabric being applied. The second coat will not require as much dope as the first coat, but a plentiful amount should be used. All coats after the second coat should be sprayed on. These coats may be brushed on but should be applied in the form of thin coats with a technique more like that used in applying a thin coat of paint or varnish. Third, fourth, and fifth coats of clear dope should be applied. Each coat should dry 30 to 45 min.

Brush coats often require thinning. A nitrate dope thinner of standard brand should be used. Thin only enough to reach the proper brushing quality. Most dope requires some thinning for spraying. Dope used with pressure-spraying equipment requires less thinning than that used with suction equipment. Use as little thinner as possible.

The fifth coat of dope, after thoroughly drying several hours, should be sanded lightly, using a No. 000 sandpaper. When sanding dry dope, the structure should be grounded to avoid sparks caused by static electricity. Clean the surface thoroughly and apply two heavy, wet coats of aluminum pigmented dope. Drying time should be 30 to 45 min. between coats.

Aluminum pigmented dope should always be used over the clear dope as a protection against destructive rays from the sun. Aluminum pigmented dope makes it possible to obtain a finer finish than clear dope. Approximately 1½ lb. of fine aluminum powder to each 5 gal. of unthinned clear dope should be used. The dry powder should be wet with dope thinner, making a thin paste before adding to the clear dope. Allow 8 hr. drying time after the second coat of aluminum dope is applied. The surface should then be sanded to a smooth finish, using No. 280 wet-or-dry sandpaper and water. This operation determines largely the final finish. Clean thoroughly, wiping off all dust, dirt, and sanding mud. A sponge and chamois dampened in clean water should be used. Apply 3 or more coats of pigmented dope by spraying. These coats should be thinned by adding 20 to 30 per cent thinner or reducer. Allow 30 to 45 min. drying time between coats. After drying 8 to 12 hr., the pigmented coats should be sanded lightly with No. 400 wet-or-dry sandpaper and water and polished with a rubbing compound.

The above-described method of applying all coats wet and using the dope as heavy as practicable should produce a high polish when a rubbed lacquer or pigmented dope finish is applied. The final coats, or pigmented dope finishes, are applied by spraying on four to six coats. Every second coat should be lightly sanded with No. 320 wet-or-dry sandpaper with water. The last coat is usually diluted with approximately ⅛ antireducer or normal reducer to prevent blushing. The final coat does not require sanding or rubbing.

Removing Dope and Lacquer from Covering. Ordinary types of paint and varnish remover have an acetone base and contain a small quantity of wax to prevent rapid evaporation. Acetone evaporates rapidly and, without wax, would not remain on the surface a sufficient length of time to remove the paint. Removers containing wax should not be used on fabric or plywood. Dope will not dry over wax. There are a number of good dope and lacquer removers on the market which are wax-free.

The complete removal of dope finish from fabric is best accomplished by placing the surface to be treated as nearly horizontal as possible. When the surface cannot be arranged in a horizontal position, soft rags may be soaked in the remover and spread over the surface to be treated. This allows the remover to remain in contact a sufficient time to loosen the dope. The dope is wiped and scraped off as the solvent takes effect. When the opposite side of the fabric from which the finish is to be removed is accessible, rags soaked with remover are placed on the old finish to be removed and allowed to dry overnight. The opposite side of the fabric is then wet with remover, thoroughly loosening the finish. The rags on the outside are then pulled off and will remove practically all of the original finish. The general method for removing finish is to treat an area approximately a foot square at a time, completely removing the finish before passing on to a new area. Remover should be applied and, as soon as the finish is softened, it should be scraped off with a wide, flexible putty knife or some instrument which will not damage the fabric. The fabric is then rubbed with rags soaked in remover to take off the remaining dope.

When small areas, because of checks and blisters, need to be refinished, a clean, soft rag wet with remover may be used. Rub the area with a circular motion, soaking and wiping the dope at the same time. While this area is still soft, build up again with the same number of coats originally used. Add 25 to 30 per cent of dope remover to the first coat. This softens the old dope and gives a better bond between the old and new.

Dope should never be applied over old surfaces which have been waxed unless the wax is thoroughly removed with a wax solvent. Oil, grease, and dirt must always be thoroughly removed from the surface before new coats of finish are applied. Old coats which are cracked or checked may sometimes be refinished by removing the outer layers and refinishing with aluminum pigmented dope and pigmented dope as in the original finish.

Plywood finishes may be removed in much the same manner as fabric finishes, scraping off the original finish with long, even strokes with a broad-blade putty knife or other suitable tool. Special finishes for gliders and light airplanes are on the market.

Plywood Finishing. When plywood is used as an outside covering, it is sometimes finished by having a coat of dopeproof material placed on the plywood which is then covered with fabric which is doped and finished with pigmented dope. Where no fabric covering is used, the plywood is thoroughly sealed with a special plywood sealer. A heavy coat is applied which should fill in all minute cracks and crevices in the plywood. This coat should be applied with a brush, keeping the brush thoroughly wet and filled with the sealer. If possible, the structure may be dipped in a tank of sealer, which is the most effective method. The least effective method is application by means of a spray. If it is necessary to use a spray, the mixture should be set as wet as possible and as heavy a coat as practicable be applied. A second and third coat of sealer should be sprayed on. After thoroughly drying, two coats of aluminum pigmented finish should be sprayed on, allowing 30 to 45 min. for drying between each coat. The plywood then may be finished with any regular pigmented plywood finish.

Finishing Steel and Metal Parts. Metal parts require rustproofing on their interiors. "Lionoil" is usually used for this purpose. Exposed metal parts and fittings should be cleaned of all rust or scale, and a zinc chromate primer with about 4 oz. of aluminum bronze powder per gallon is used. Two or more coats are applied.

Areas of Contact between Wood and Metal. All interior surfaces of contact between wood and metal fittings should be marked off and painted by brushing with a generous coat of bituminous paint to which aluminum paste has been added. The estimated spreading rate is 300 sq. ft. per gal., weight 0.018 lb. per sq. ft., drying time 8 hr., and moisture-excluding effectiveness 80 to 90 per cent. Occasional practice, however, is to use an additional coat of phenol formaldehyde varnish instead of bituminous paint on metal-to-wood contacts.

XIV GENERAL SAFETY RULES

Most tools in the hands of unskilled workmen are apt to be dangerous and may cause serious and painful injuries. Although an injury may be slight, if the skin is broken first aid should be given. Every injury, no matter how slight, should be reported to the instructor or person in charge of the shop. Most injuries heal rapidly unless infected, and every precaution should be taken to keep the wound clean. Most tools used in any shop are not free from germs. Particles of wood, metal, dirt, oil, etc., can be carried into even a slight wound made by tools. No matter what the tool is that you are using, get in the habit of using it in such a manner that, if it slips or gives way, the tool will not injure you and you will not be off-balanced and injure yourself by falling against benches, machinery, or the tool itself. A fundamental rule in all shops or laboratories is never to touch anything with which you are not entirely familiar and have not received instruction in the proper method for its use. This applies not only to machinery and various tools, but also to the other person's job. Do not pick up and examine the other person's work or his tools. Do not talk to other persons while they are engaged in their work. Do not borrow tools, and it is good practice to refuse to lend your own. Practical jokes in a shop may turn out to be tragedies. Do not lift weights that are beyond your strength. In lifting heavy weights, place your feet close together, get a firm hold on the work, keep the back vertical, and lift slowly with the leg muscles which are the most powerful in the body.

Do not leave your tools lying around. Sharp-edged tools lying over the edge of a bench may cause serious injuries. In working on ladders or high places do not put tools down where there is danger of their falling. They may strike another person or injure the tool itself or harm some finely finished piece of work. Never run in a shop, as you may fall and receive serious injury from operating machinery or sharp tools. When carrying sharp tools, have the edges covered with a leather guard or wrapped in other heavy material. Above all, do not run with sharpened tools in your hand. Do not wear long neckties, long sleeves, or any clothing which may be caught by whirling machinery.

Follow shop rules exactly. If shop rules require that you wear goggles, safety shoes, gloves, etc., on the job, do so. There are many power tools used in the modern shop, and human flesh and bone is more easily cut than wood and metal. The band saw is a common piece of equipment in woodworking shops and, while the blade is usually very well protected, all precautions in using it should be observed. The blade moves very rapidly, and slight contact with it results in a severe injury. If the work requires the use of the hands in close proximity to the blade, use a wooden block to push the work past the blade. If the blade appears to be cracked or clicks in passing through the work, cut off the power and wait until the blade has completely stopped before trying to remove the work. Band saws should not be used, except when absolutely necessary, in temperatures at, or below, freezing. Most metals break easily when cold. In shutting off the power do not try to stop the blade, let it coast to a standstill. In using a circular saw, always push the work into the saw with a properly notched wooden block. Do not use the hands to push the work past the saw. Always stand slightly to one side so that if a splinter or piece of wood is thrown away from the saw it will go past you and not through you. Be careful that a piece of wood is not picked up by the back of the saw and thrown over the blade toward you. Be careful never to reach between belts or over moving parts of machinery to pick up objects or move work. Walk around the machine or stop it before reaching through. Many shafts have a collar and a set screw on the end which may become entangled with clothing and twist the operator into the machinery.

In using sanding machines, care should be taken that the hands do not come in contact with the sanding belts or wheels. Do not try to remove too much material at one time. Use light pressure on the work.

Power tools are not the only dangerous tools in the shop. Each hand tool requires proper use if the operator is to avoid injuring himself or his work.

Hammers are designed for a particular use. Do not try to use hammers which are not designed for the work for which they are being used. One danger in using a hammer is its head flying off. The head may fly off at high speed and is almost as effective as a bullet. Keep the hammer handle free from grease. Be sure that you have a good grip on the handle when using the hammer. Look where you are going to strike. Do not try to watch the hammer head. In driving nails there is danger of the nail flying away from the work with considerable force. Start the nail well before giving it a firm blow. Be sure the face of the hammer is free from grease or dirt. Be sure the face of the hammer meets the work squarely. Do not use anything but the face of a hammer in pounding. Do not strike metal that is harder than the face of the hammer. This will cause severe injury to the hammer face and prevent effective use later.

A sharp tool is more effective to work with than a dull one. Many injuries are caused by striking the sharpened edge of the tool. Always put sharp-edged tools away when not in use.

A machinist once remarked that he had "lost so much skin from his knuckles during his apprenticeship, that if it could all be placed in one piece and tanned he could make himself a leather apron."

Many knuckles are skinned by having a wrench slip off a nut. Whenever using wrenches, be sure that the jaws are in good condition and the wrench is placed on the nut so that the pressure exerted on the handle will tighten the jaws. Before placing pressure on the handle of a wrench, think about what will happen to your knuckles if the wrench should slip. Never use an extension to lengthen the handle of the wrench. The handle of each wrench is designed long enough to exert all the pressure the wrench will stand in the workman's hand. Whenever possible, pull on a wrench handle instead of pushing. When standing on a platform or a ladder, be sure that you are so braced that if the wrench slips you will not fall.

The screw driver is a simple tool which is used as often as any tool in your kit. A screw driver should never be used for any other purpose except for removing or setting screws. In starting a screw, use a small auger, awl, or drill to make a hole in which the screw will start. Tap the screw lightly into place. Be sure the screw driver fits the slot perfectly before attempting to set the screw. Do not set a screw too firmly as this may break the head or tear the wood fibers. If the blade of the screw driver becomes nicked or bent, always reshape it with a file or grinding wheel before using. Do not hold a screw with the fingers while you put pressure on it with the screw driver. The screw may slip from under the screw driver and the sharp blade may cause injury to the hands. Do not use a screw driver for a punch, chisel, or to pry off blocks of wood. Use it only for its intended purpose.

Files can cause injury. The sharp tang should always be covered with a good file handle. It is sometimes necessary to file rotating material. Be sure that the file is held firmly and not allowed to be twisted out of your hands into the machine.

Wood chisels have keen edges, and the manner in which they are used makes them likely to slip. Do not use the chisel toward any part of your body. Look ahead and see what would happen if the chisel were to slip. Be sure the chisels are kept properly sharpened and placed in their proper containers after use. If the edge has become dull always sharpen it before putting it away.

Saw teeth are sharp and are apt to cause painful injury. In sawing, be careful that the saw does not jump out of the groove and come in contact with the hand holding the work.

There is not much danger in using planes of various types, but in sharpening the blades care should be taken not to cut the hands. Do not hold the stone upon which you are sharpening the blade in one hand and try to sharpen with the other. Place the stone in its proper box or brace it against a bench stop or other firm block while using.

When using a draw knife, do not brace the work with your knee or any part of the body.

In sharpening tools be sure that the grindstone, grinding wheel, or sharpening stone is clean. In using a steel or a whetstone to sharpen knives, be sure that your hand is protected with a guard of heavy leather, rubber, or soft metal. A grindstone should always be turned away from you. The tool should be held in one hand and guided, and held firmly against the stone with the other hand. In using high-speed grinding wheels always stand to one side of the wheel. Wheels sometimes break when rotating at high speed, and the piece may fly off at great velocity. Do not sharpen tools or grind material on any part of the wheel except that designed for grinding. Smooth, greasy oil stones may be cleaned by placing in a dry pan and heating. Wipe the stone with burlap or other coarse cloth while still hot, taking care not to burn the hands. In sharpening tools with a file, make the stroke away from the edge and not toward it. Use every care possible to increase safety in the shop. A lost eye or lost finger can never be replaced — and an injured workman is not an effective workman, even though he may remain on the job.

XV CONSTRUCTION OF A ROOT BAY OF A FABRIC-COVERED WING

The problems in this chapter will cover the complete construction, alignment, covering, and finishing of one bay of a fabric-covered airplane wing.

The bay of a wing is that portion lying between successive compression members. Each bay is a complete unit with the exception of the spars which run the entire length of the wing. There are two spars — a front spar and a rear spar. Each bay is self-supporting and forms the successive units of support in the wing.

The root bay consists of the following parts: front spar, rear spar, compression members, glue blocks, wood braces, fittings, wing covering, form ribs, false ribs, leading-edge strip, leading-edge cover, trailing edge, and drag and antidrag wires.

The root bay to be constructed in the following problems includes the different types of construction which may be found in any standard wing.

A number of different ribs will be constructed so that the woodworker will become familiar with various types of ribs. All construction is standard.

The bay will be 38¼ in. in length from the root of the spar to the center of the fittings to which the drag and antidrag wires are fastened. The first rib will consist of a root or walk rib. This rib is also called a box rib. It will be covered with plywood and have a walk strip covered with abrasive cloth. The walk or box rib consists of the root compression rib and a reinforced rib covered with a plywood walk.

In addition to the root or box rib, there is a form rib, a reinforced compression rib, and two false ribs. The compression rib must support the strain placed on it by the drag and antidrag wires. The false ribs are the short ribs at the leading edge of the wing which maintain the shape of the leading edge between the main or form ribs.

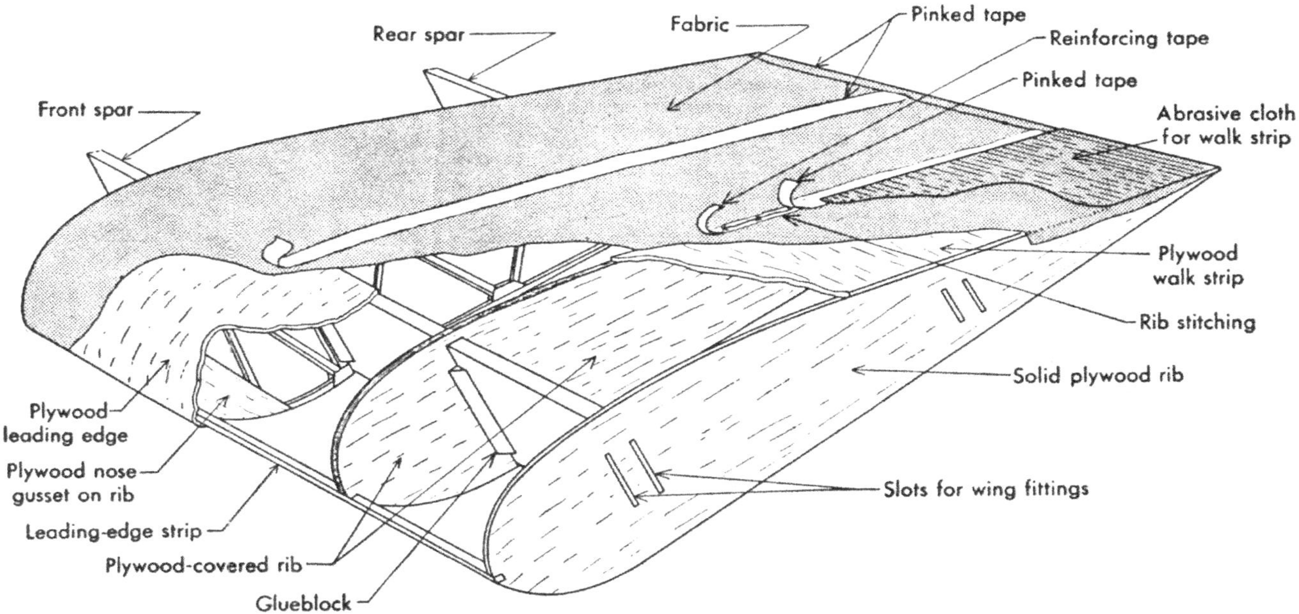

Fig. 140. Sketch showing the construction and parts of a root bay of a fabric-covered wooden wing. (Not a scale drawing.) Root of right wing.

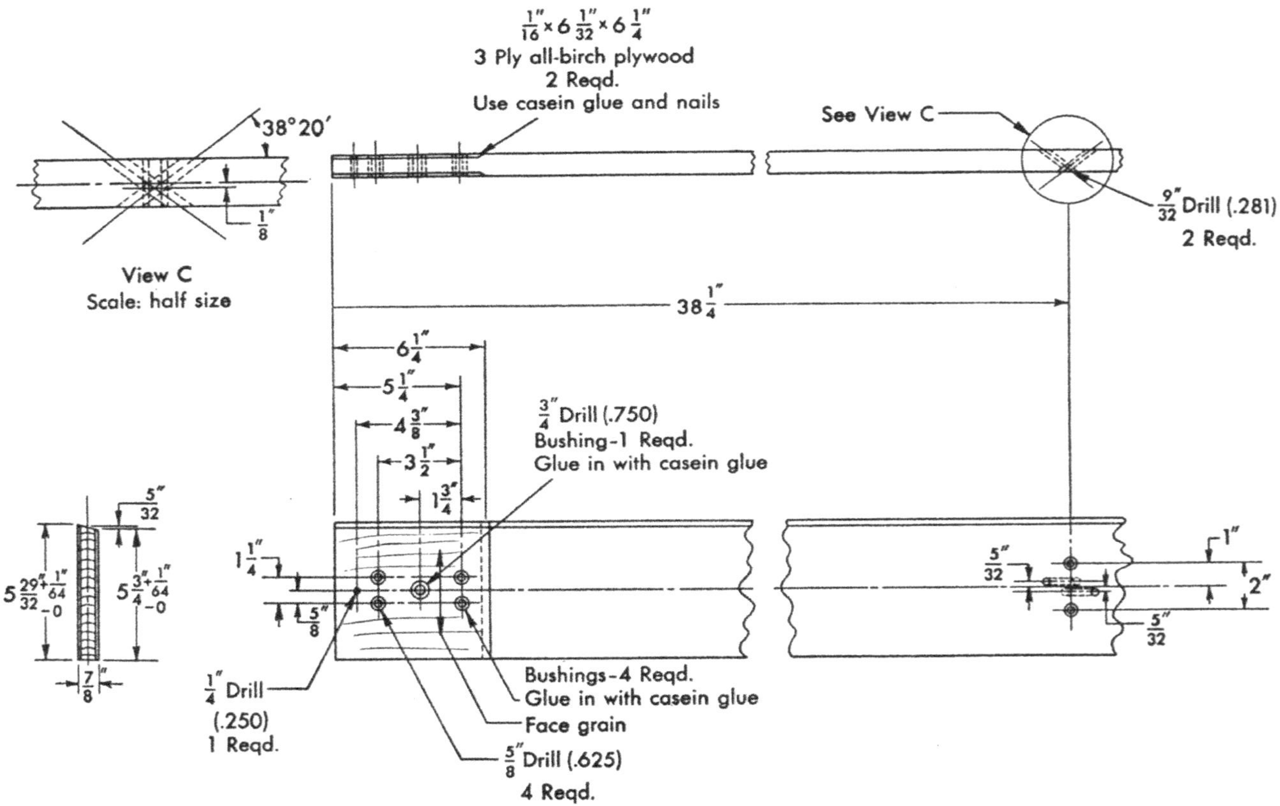

Fig. 141. The construction of the root portion of the left front spar.

Details of the fittings will be given. The wires and fittings may be made from standard materials and fabricated in the shop. These fittings, since they will only be used in this course to align the wing and will not be subjected to the strains imposed upon fittings in a completely constructed airplane, may be made of any material which is strong enough to hold the bay in shape for assembling and covering. It may be desirable to build temporary fittings of wood.

Except for the experience gained in identifying the proper grades of material to be used, it is not necessary that the material used for the spars be Grade A aircraft spruce. It is desirable, however, that the proper grade of material be used whenever possible.

Problem 1. Construct the Root-Bay Section of the Front Spar. For the front spar, a spruce plank 6½ in. wide, 1 in. thick, and not less than 48 in. long will be needed. Figure 141 gives the dimensions of the cross section of the spar, and these dimensions should be followed exactly. The piece of spruce plank should comply with the specifications given in Chapter V. Plane one side and edge of the plank until smooth and at right angles. Care should be taken that this edge is exactly straight, as it will be the bottom edge of the spar. Carefully square one end of the plank which will be the root end of the spar. The straightness may be tested by a steel straightedge. The right angle may be tested with a try square. The second side of the spar should be dressed down until the spar is ⅞ in. thick. Bevel the top of the spar. Care should be taken that this bevel is exact. The depth of the front face of the spar is 5¾ in., and the depth of the rear face is $5\frac{29}{32}$ in., as shown in Figure 142.

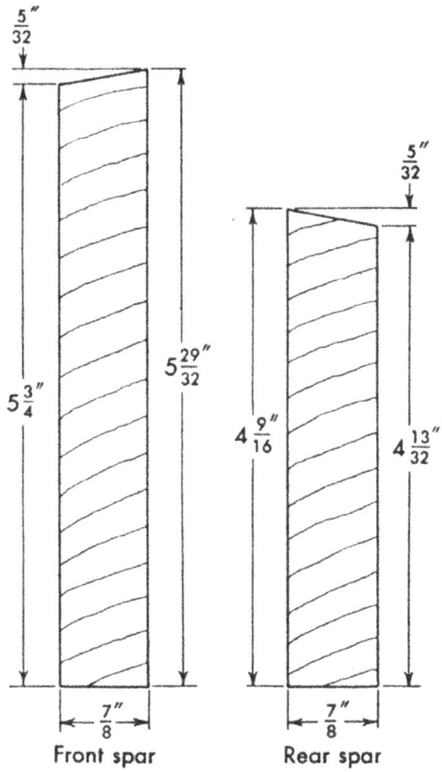

Fig. 142. Cross section of a front and a rear spar.

It is good practice, whenever fittings are to be bolted to spars or other structural members, to reinforce these points with plywood. These pieces, as shown in Figure 141, extend 6¼ in. from the root of the spar and from the bottom edge of the spar to the beveled edge. The pieces should be beveled to match the bevel on the top edges of the spar after they are in place. To install these plywood reinforcings, a line should be drawn across each face of the spar parallel to the root end of the spar, exactly 6⅛ in. from the end of the spar. A saw cut 1/16 in. deep should be made, and 1/16 in. of the spar material should be carefully removed by the use of a broad chisel or a block plane. When this material has been removed to the proper depth, the square corner left by the saw cut should be beveled as shown in Figure 141. The plywood pieces which are to fit on each side of the spar should have one edge beveled to the same slope as that part of the spar just beveled. This joint should be as nearly perfect as possible. The plywood pieces are fastened in place by the use of casein glue and nails. These nails should be aircraft nails and enter the spar at least 3 times the thickness of the plywood. Plywood should be prepared for both sides of the spar and glued at the same time.

Proper gluing blocks should be applied. These consist of hardwood blocks having a smooth surface and a thickness of not less than the thickness of the spar and a width slightly in excess of the width of the spar. C-clamps should be used to exert pressure. The plywood should be lightly sanded to remove the finish and give a better gluing surface. The proper amount of casein glue should be prepared and spread evenly over both the plywood and the surface of the spar to which the plywood is to be glued. The plywood on the rear of the spar should be placed in position and nailed with staggered rows of nails approximately 1 in. apart. Nails should not be placed where holes are to be drilled. Plywood is applied to the front face of the spar in the same manner. Before placing the gluing blocks on each side of the spar, cover the plywood and the glued edges with a sheet of paper to prevent the glue sticking to the glue block and ruining the plywood when the glue block is removed. Place C-clamps at about 1½ in. from each corner of the root of the spar and two clamps over the glued edge of the plywood. Tighten the clamps as tightly as possible by hand. A pressure of approximately 150 lb. per sq. in. should be exerted. After about 10 min. the clamps should be retightened because the glue being forced into the wood relieves some of the pressure. The glue should be allowed to set not less than 6 or 8 hr., preferably overnight. The clamps are then taken off, and the paper and excess glue carefully removed by means of a scraper or sandpaper.

The drilling of holes for aircraft fittings requires a high degree of skill. Any variation in drilling will place undue strain on surrounding parts of the wood. The proper method for drilling the fitting holes at the root of the spar is by use of a hardwood jig.

Most parts of an aircraft are built with the aid of a jig. A jig is any rigid structure or mechanism of metal or wood or other material which holds parts being fabricated while they are glued, welded, sawed, drilled, etc., before assembly or while being assembled. A jig holds the various parts of the structure while it is being built up or torn down. Parts such as ribs which must duplicate each other are built in this manner.

A piece of hardwood approximately 1½ in. thick, the width of the spar, and approximately 8 in. long should be used to prepare the jig for drilling these holes. This block should be smooth on all sides. One end and one side should be finished to fit exactly along the end and bottom edge of the spar. The holes shown in the root of the spar should be carefully laid out on this jig and drilled on a drill press. The proper sizes of wood drills to be used are shown in Figure 141. When drilling the holes in the jig, a backing block of material thick enough to prevent splintering when the drill breaks through the jig should be used. After the holes are drilled in the jig, they should be carefully checked to see that they are true and in the proper places. To drill the holes in the spar, the jig should be placed on the rear side of the spar so that the finished square end of the jig is even with the end of the spar and the edge of the jig is even with the bottom edge of the spar. A backing block of not less than 1-in. material should be placed on the opposite side of the spar and both the jig and the backing block clamped to the spar by the use of C-clamps. After clamping, the jig should be carefully checked to see that it is in the correct position. When the jig is in place, drill the holes in the spar.

At the outer end of the bay four holes are to be drilled for the drag and antidrag wire fittings. Note that two of these holes are drilled at right angles through the spar and are fitted with bushings. The bushings may be ordered, or made from any suitable material. It is suggested that temporary bushings be made from soft materials such as brass, plastic, or hardwood. A jig should be prepared in the same manner as for drilling the holes at the root of the spar. Two holes are drilled at an angle of 38° 20′ through the spar. A jig should be prepared from a block of wood and drilled as shown in Figure 141. In drilling these holes, a suitable backing block should be clamped on the opposite side of the

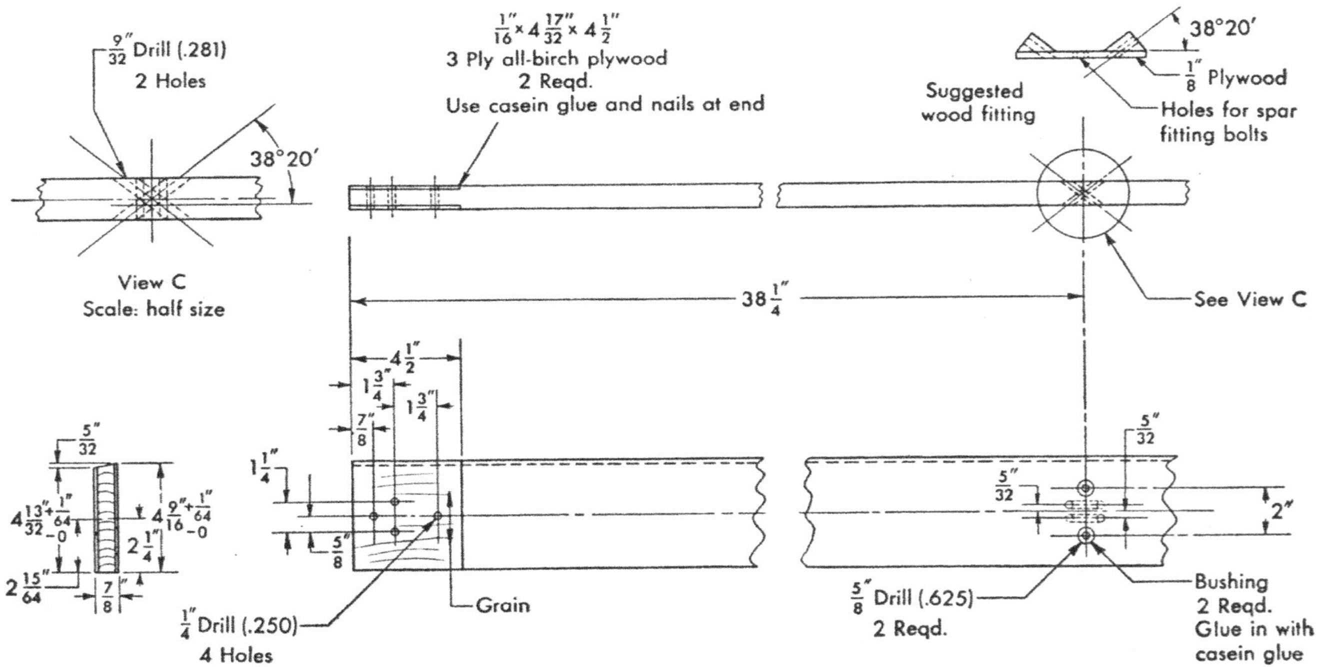

Fig. 143. The construction of the root-bay portion of the lower left rear spar.

spar from the jig. The jig should be so marked that the holes will be in the proper place.

Problem 2. Construct the Rear Spar. The rear spar is constructed in the same manner as the front spar except for the difference in the dimensions. The rear spar is $\frac{7}{8}$ in. thick, the front face is $4\frac{9}{16}$ in. in depth, and the rear face is $4\frac{13}{32}$ in. in depth. Care should be taken that the bevel on the rear spar is made in the opposite direction from that on the front spar. The narrow face of the front spar is to the front, and the narrow face of the rear spar is to the rear.

Problem 3. Construct a Rib Jig. A jig must be prepared for fabricating the ribs in this problem. Select a piece of lumber, preferably hardwood, of the following dimensions: 14 in. wide, 72 in. long, and approximately 1 in. thick. This piece will be used as the base of the rib jig. Draw on this board a straight line 1 in. from one edge of the board. This will be the base or reference line from which the rib will be laid out. At a point 4 in. from the end of the board, where the nose of the rib will be laid out, erect a perpendicular line to the base or reference line; and $58\frac{1}{2}$ in. from this line erect another perpendicular to the reference line. Along the reference line, lay off the dimensions given in Figure 144, the first distance being $\frac{31}{32}$ in. which is the radius of the nose of the rib, then $2\frac{1}{2}$ in., etc., until all of the dimensions given along the reference line are clearly marked. These points are called "stations." Note that each dimension given is the over-all dimension from the nose of the rib, the total length of the rib being $58\frac{1}{2}$ in. Erect a perpendicular at each station on the reference line. On these perpendiculars lay off the dimensions, beginning with the $\frac{31}{32}$ in. station. On the perpendicular at this station, lay off two dimensions, one $3\frac{1}{8}$ in. from the reference line, and the other $5\frac{5}{8}$ in. from the reference line. On the perpendicular at the $2\frac{1}{2}$ in. station, lay off $2\frac{19}{32}$ in. and $6\frac{9}{16}$ in.; at the $5\frac{1}{16}$ in. station lay off $2\frac{1}{4}$ in. and $7\frac{17}{32}$ in. on the same perpendicular. At the 9 in. station lay off 2 in. and $8\frac{7}{16}$ in. on the same perpendicular. The remainder of the length of the bottom of the rib is a straight line which will be 2 in. from the reference line. From this point to the trailing edge of the rib only one dimension will be laid off on each perpendicular. At the nose of the rib on the perpendicular erected at the $\frac{31}{32}$ in. station, a point $4\frac{9}{32}$ in. from the base line is marked, which is the center of the $\frac{31}{32}$ in. radius, to form the nose of the rib.

After all dimensions given are marked on the jig base with a flexible piece of wood or plastic, the points outlining the top and bottom of the

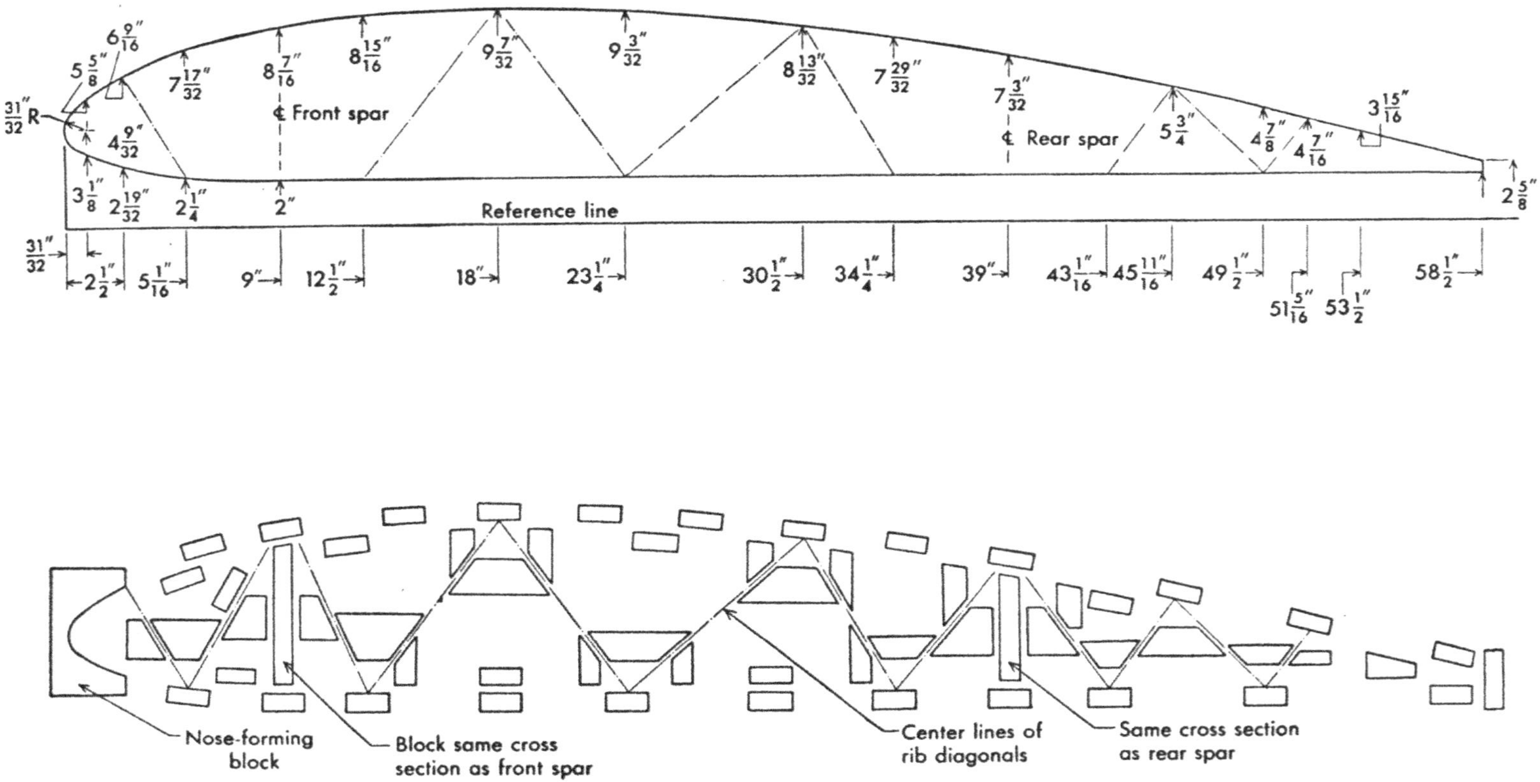

Fig. 144. A layout of a rib jig.

rib are drawn in as shown in Figure 144. After the rib has been outlined, suitable blocks are prepared from either 3/8 in. hardwood or 3/8 in. plywood and placed around the outline of the rib, as shown in Figure 144. These blocks should be glued and nailed or screwed into place. The block forming the outline of the nose rib should be carefully prepared and glued and screwed into place to form the nose portion of the jig. After the outside blocks are in place and allowed to dry, the measurements of the various stations should be carefully checked. Blocks around the cambered portion of the rib should have the edge next to the line shaped just to touch the line throughout the length of the block.

Prepare a nose block and place in the jig. The block should fit snugly against the nose-forming block and be nailed temporarily into place.

Cap strips for both the top and bottom of the rib should be prepared. This strip should be 3/8 in. square and of straight-grained spruce such as will be used to build the rib. These strips should be slightly longer than the top and bottom cap strips of the ribs. Place these strips in their proper place as though they were top and bottom cap strips, holding them in place by temporary nailing. These strips should fit closely against the blocks forming the outer outline of the top and bottom cap strip. Place the blocks inside the cap strip, as shown in Figure 144, forcing them snugly against the temporary cap strips, and nail and glue. After these blocks are in place, prepare two blocks which are the same in size as the cross section of the front and rear spar. Fasten these blocks in place with glue and screws.

Prepare rib cross members as shown in Figure 145. Place one on each side of the blocks representing the spars. These pieces will be the same as the verticals in the rib on each side of the spar. Place diagonals of 3/8 in. square spruce and fasten temporarily along the center lines as shown in Figure 144. A temporary rib is now in place in the jig. Blocks should be prepared as shown in Figure 144 to fit snugly against these diagonals. These blocks are to be glued and nailed or screwed into place. The jig is complete and should be allowed to dry thoroughly, and the pieces forming the rib removed. The entire jig should be coated with hot paraffin to prevent the rib from sticking to the jig. All dimensions should be carefully checked to see that they agree with the drawing.

Problem 4. Construct a Rib. The different types of ribs will all use the same jig for their main framework. Before placing the cap strips in the jig, prepare a nose block of 3/8 in. basswood as shown in Figure 145. Use 3/8 in. square spruce throughout for cap strips and diagonals for

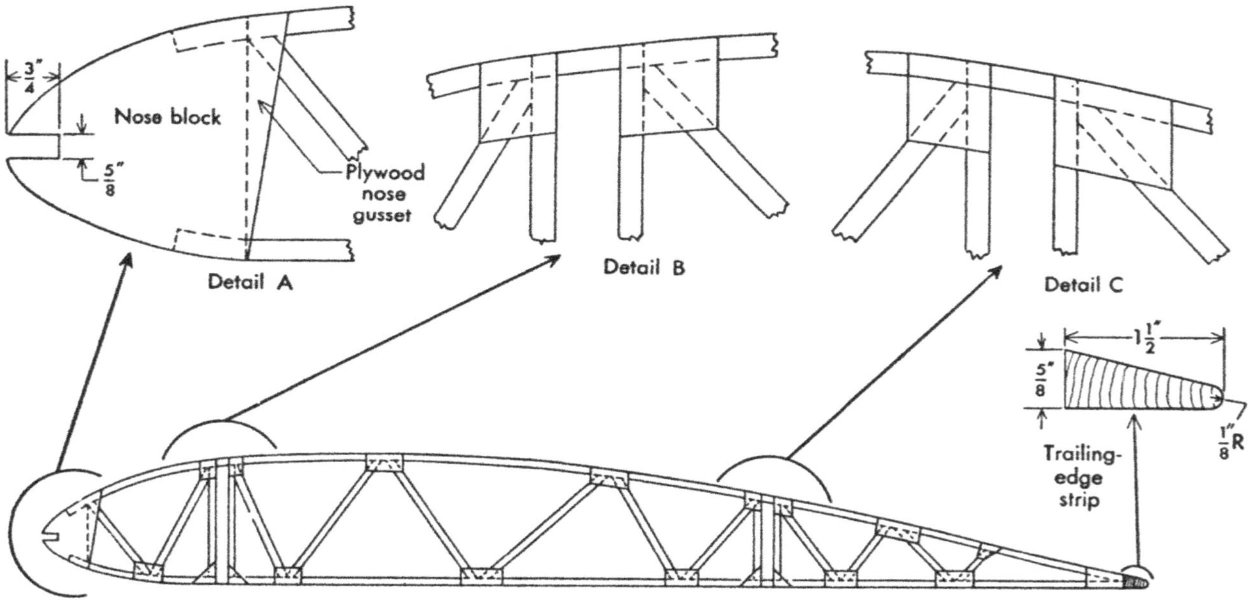

Fig. 145. The construction of a main rib. Rib No. 3 in the root bay.

all ribs. Top and bottom cap strips are first placed in position in the jig. Gussets are prepared of 3/32 in. plywood, as shown in Figure 145. Preferably, this plywood should be mahogany with a cottonwood or basswood core. Be sure that all pieces fit snugly in the jig. Coat the gussets with prepared casein glue and nail into place, using flat-head aircraft nails which penetrate the rib material 3 times the thickness of the gusset. A plywood piece for the nose of the rib should be prepared of the same material as the gussets and glued and nailed into place. Allow the rib to dry not less than 6 hr. Remove the rib from the jig, turn it over on a flat surface, and nail and glue duplicate gussets on the opposite side of the rib. After drying, the rib is complete.

Problem 5. Construct the Root Rib. The root rib is formed of solid 3/8 in. plywood. For this problem any suitable plywood may be used but should not be made up of less than 5 plies. This rib is solid plywood, notched at the nose to receive the leading-edge strip. This rib is not cut out to receive the root of the spar but would normally have openings cut through it for the spar fittings. The rib is fastened to the end of the spars with glue blocks as shown in Figure 146. This root rib acts as a compression rib.

Problem 6. Construct Reinforced Rib No. 2. Place in the jig a top and bottom cap strip and nose piece. Verticals on each side of the spars are placed the same as for a main rib. The other cross pieces are not used in this rib. This rib will be covered on both sides with 1/16 in. plywood. The plywood pieces to cover each side of this rib must have openings cut the same size as the cross section of the front and rear spars shown in Figure 142. All portions of the rib in the jig are coated with casein glue and the plywood webs are placed in position and nailed with the same-sized nails used for the gussets in the main rib. Allow the plywood to dry not less than 6 hr., remove the rib from the jig, placing it on a flat surface, plywood side down, and place the stiffeners in position as shown in Figure 147. The stiffeners should be coated with glue on both sides. The entire exposed portion of the rib to which the plywood will be fastened is also coated with glue. Place the plywood in position, as shown in Figure 140, and nail. Allow the rib to dry not less than 6 hr.

Problem 7. Construct Compression Rib No. 4. Rib No. 4 will have a plywood web and a double top and bottom cap strip with reinforcing verticals. This rib is formed by using a regular top and bottom cap strip, a nose piece, and four verticals. A second half of the rib is built up like the first half, except that the gussets are placed on this half while in the jig. After removing from the jig, the side of the rib opposite the gussets is covered with casein glue and nailed to the plywood web on the first half, using nails which will almost penetrate both portions of the rib. Plywood gussets are now nailed and glued to the opposite side of the rib.

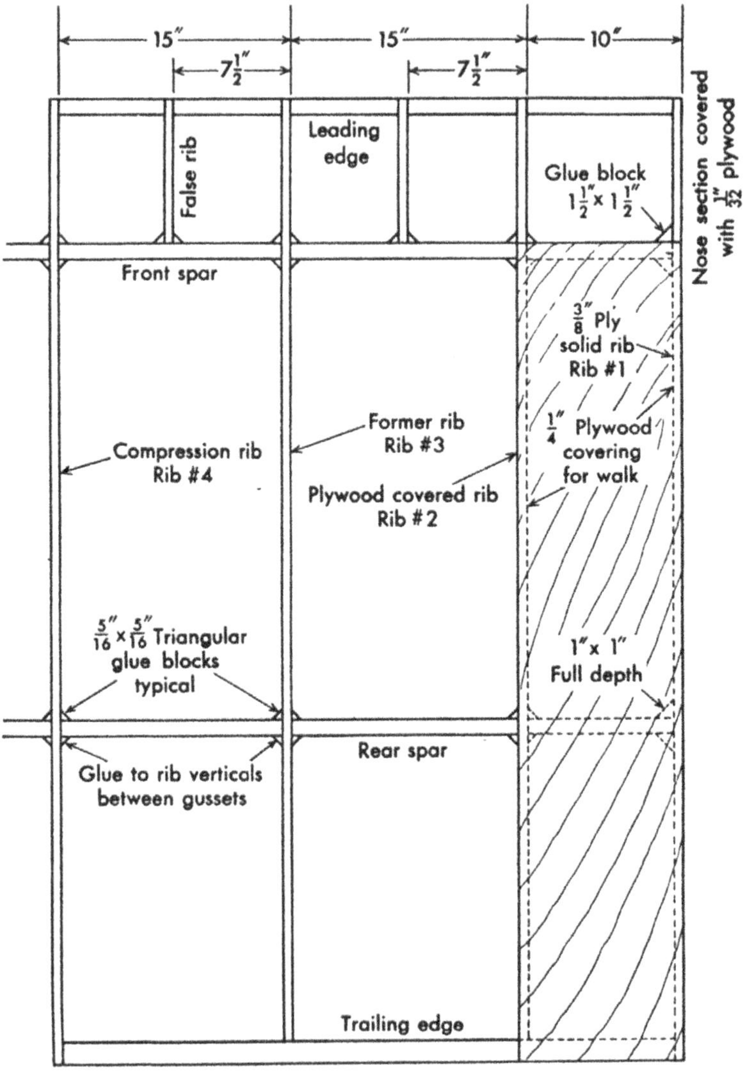

Fig. 146. A plan drawing showing the main structure of a root bay.

Problem 8. Construct the False Ribs. False ribs are short ribs which help to support the leading edge of the wing and extend only from the front spar forward. Two false ribs are to be constructed. A top and bottom cap strip extending from the nose approximately 2 in. back of the front

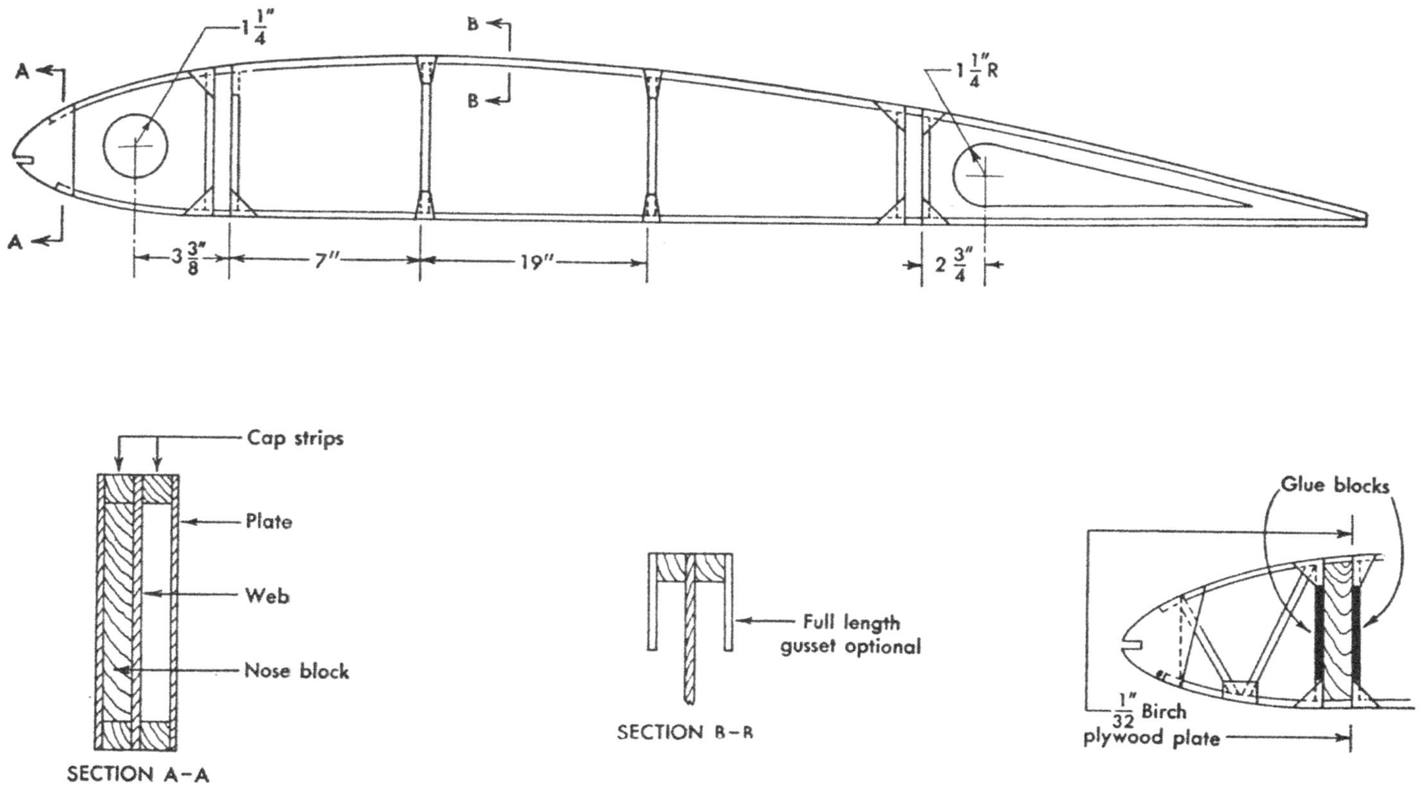

Fig. 147. A double rib with vertical stiffeners and full plywood web. Rib No. 4 in the root bay.

spar is placed in the jig. The rib is then formed in the same way as a regular rib, having the two verticals between which the front spar passes and a nose portion identical to the main ribs.

Problem 9. Construct the Leading-Edge Strip. The leading-edge strip is constructed of spruce. This strip is 40⅜ in. in length, ½ in. thick, and ¾ in. wide. The front edge is rounded off, having a radius of $^{31}\!/_{32}$ in.

Problem 10. Construct the Trailing-Edge Strip. This strip is constructed of straight-grained spruce 40⅜ in. long with a cross section as shown in Figure 145.

Problem 11. Assemble the Wing Bay. With all fittings, braces, and glue blocks complete, the root bay of the wing should be assembled as follows. The two spars are placed on supports, and ribs No. 2, 3, and 4, and the two false ribs are put into place. The fittings and wires are now assembled and tightened just enough to hold the spar firmly. The spars should be exactly parallel to each other and the ribs at right angles to the spars. The distances used to line up the assembly may be measured with a trammel gauge. When properly aligned, the glue blocks, braces, etc., are glued and nailed into place. The rib verticals on each side of the spar should be glued and nailed in place. After all the parts are glued and fastened, the contour of the nose portion of the ribs should be checked by using a straightedge. The nose strip which has been prepared is now fastened into place by the use of casein glue and nails.

Problem 12. Cover the Leading Edge. The leading edge of the wing is covered with $^{3}\!/_{32}$ in. plywood, preferably mahogany with a cottonwood core. The plywood extends from the rear edge of the top of the front spar around the nose to the rear edge of the bottom of the front spar. The proper-sized sheet of plywood should be prepared. It may be necessary to dampen slightly the outside of the plywood by covering it with wet rags or burlap for approximately ½ hr. This plywood will be glued and nailed into place. Proper nailing blocks should already be in place on the top and bottom of the spar. The plywood should be first nailed and glued to the nailing block along the rear edge of the top spar. The edges of the nose portion of the ribs and the blocks on the rear spar should be covered with glue, and the plywood bent around the leading edge and held in place. The plywood is now nailed around the leading edge of the rib, using nails which penetrate the rib not less than 3 times the thickness of the plywood, and similarly nailed along the rear edge of the bottom of the front spar. The nails should be approximately 1 in. apart. The leading edge should be allowed to dry not less than 6 hr. before further work is done on this portion of the wing.

Problem 13. Construct the Wing Walk. The wing walk, which covers the root of the wing, is the portion upon which a person steps when entering and leaving the cockpit. This walk extends from the front spar to the trailing edge and from the root rib to rib No. 2. A strip of ¼ in. five-ply mahogany-faced plywood is glued and nailed to ribs No. 1 and No. 2, as shown in Figure 140.

Problem 14. Cover the Root Bay. The root bay should be covered with standard airplane fabric. It will be necessary to use two strips. The seam

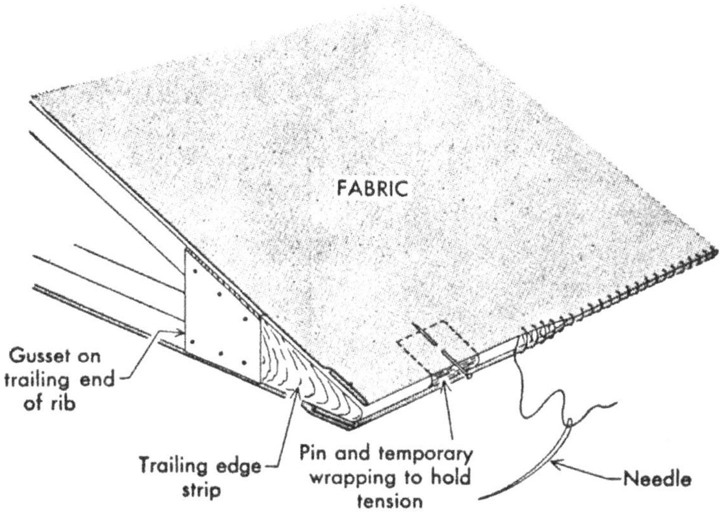

Fig. 148. A method of sewing the fabric at the trailing edge of a wing.

should run from front to back. At the root end of the wing bay, 2 in. should be allowed for lapping. At the outer end of the bay, sufficient fabric should be allowed so that it may be fastened together. The fabric beginning at the trailing edge should be carried around the leading edge and back to the trailing edge, allowing sufficient fabric for folding and sewing, as shown in Figure 148. Fabric may be tacked temporarily to the trailing edge while being sewed. The seam should be fastened at intervals of not more than 6 in. Stitches should be not more than ¼ in. apart. The thread used for sewing the seam on the trailing edge should be No. 8, four-ply cotton, having a tensile strength of not less than 14 lb. This thread should be lightly waxed with beeswax before using. The fabric should be fastened at the root end of the bay by tacking into place and will be doped to the root rib. To lace the fabric to the ribs, a strip of reinforcing tape ½ in. wide should be placed over each rib on top of the fabric. The cord used for lacing the fabric to the ribs may be cotton or linen, having a tensile strength of 80 lb., and should

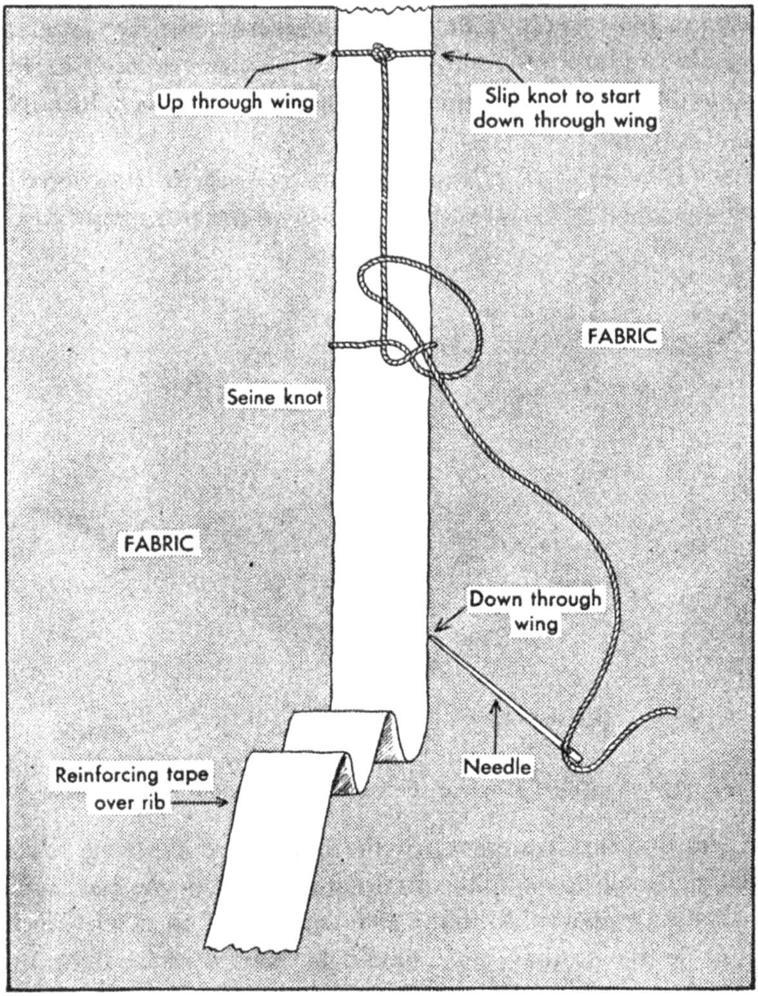

Fig. 149. A method of stitching the fabric to the ribs and the seine knot used to fasten each stitch.

be lightly waxed before using. To lace the fabric to the rib, a lacing needle of sufficient length to go completely through the wing at its greatest depth is used. The needle carrying the lacing cord is passed downward through the wing as close to the rib as possible, then back through the wing from the bottom to the top on the opposite side of the rib as close to the rib as possible. The cord is then fastened by means of a seine knot. This process is repeated the length of the rib, beginning 4 in. from the front spar. The lacing intervals should be 4 in., each lacing fastened with a seine knot, and the thread carried from one lacing to the next. The wing is now ready for doping and finishing.

Three drain grommets should be placed under the trailing edge to allow drainage of any moisture which may accumulate within the wing. Pinked tape of the same material as the wing covering will be used to cover all seams, trailing edges, and ribs. This tape may have either a pinked, scalloped, or frayed edge.

The dope to be used is a clear nitrate dope, applied with a fairly large brush. The fabric covering will be finished as directed in Chapter XIII.

Problem 15. Apply the Wing-Walk Tread. The walk rib constructed in the root bay should be covered with an abrasive cloth to prevent slipping when a person steps on the wing, and to protect the fabric from wear. This abrasive cloth is covered with small-sized grit and can be obtained from any aircraft supply house. The material should be cut to fit along the edge of the panel and extend outward to the edge of the wing walk. Pin the material from about 2 in. below the leading edge of the wing over the top of the wing and allow it to lap about ½ in. over the trailing edge. With a pencil, mark a guide line across the wing where the outer edge of the wing-walk tread is to be placed. Apply a strip of masking tape along this line on the wing top, away from the wing walk. With fine sandpaper, thoroughly sand the fabric which already covers the walk rib. Apply a coat of elastic cement to the underside of the abrasive cloth and to the area of the wing to be covered. After the first coat has dried, spread a second coat of the cement on the abrasive cloth. Allow this coat to set for 5 or 10 min. until it is no longer sticky. Apply a small amount of thinner with a brush to the surface of the cloth covered with cement and immediately apply the cloth to the area to be covered. Apply it first over the leading edge. Use a hand roller and roll the cloth firmly into place. The trailing edge should be fastened in place by a thin metal strip which prevents damage to the thin trailing edge of the wing.

XVI CONSTRUCTION OF A PLYWOOD-COVERED WING-TIP BAY

The construction of a wing tip involves several problems which are not encountered in the construction of the root bay or other bays of a wing. The tip to be constructed in this problem is to be covered with plywood which will require somewhat different internal construction.

Problem 1. Construct the Spars. The front spar will be 48 in. long, the rear spar 53 in. long. The spars will have the same cross section as the spar in the root bay. The front spar will be tapered the last 15½ in. of its length as shown in Figure 150. The rear spar will be tapered for 20½ in. as shown in the same figure. The wing-tip bow should be constructed by laminating ⅛-in. strips of white ash.

Problem 2. Construct the Wing-Tip Bow. The wing-tip bow is made of sufficient length to form a 7½-in. scarf joint on both leading and trailing edges. The leading-edge strip should be made of clear white pine and is ⅝ in. by ¾ in., edge grain on the wide face. A jig should be prepared for the lamination of the wing tip as shown in Figure 151. The curved part of the wing-tip bow on both the front and rear ends curves with an outside radius of 20 in. From the point of tangency of these arcs, the remainder of the bow curves with an outside radius of 37 in. The laminated strips should be approximately 128 in. in length as the total length of the bow when completed along the outer edge will be approximately 118 in. As shown in Figure 151, semicircular blocks of 1-in. hardwood are sawed, having an outside radius of 18½ in. to form the two ends of the bow. These blocks should be approximately 20 in. in length. A straight block 7½ in. long is cut for each end of the bow, and the rest of the blocks are cut with an outside radius of 35½ in. to form the center part of the bow. Prepare two outside end blocks for the straight ends of the bow approximately 8 in. long; four blocks, each 6 in. long with an inside radius of 20 in.; and six blocks 6 in. long with

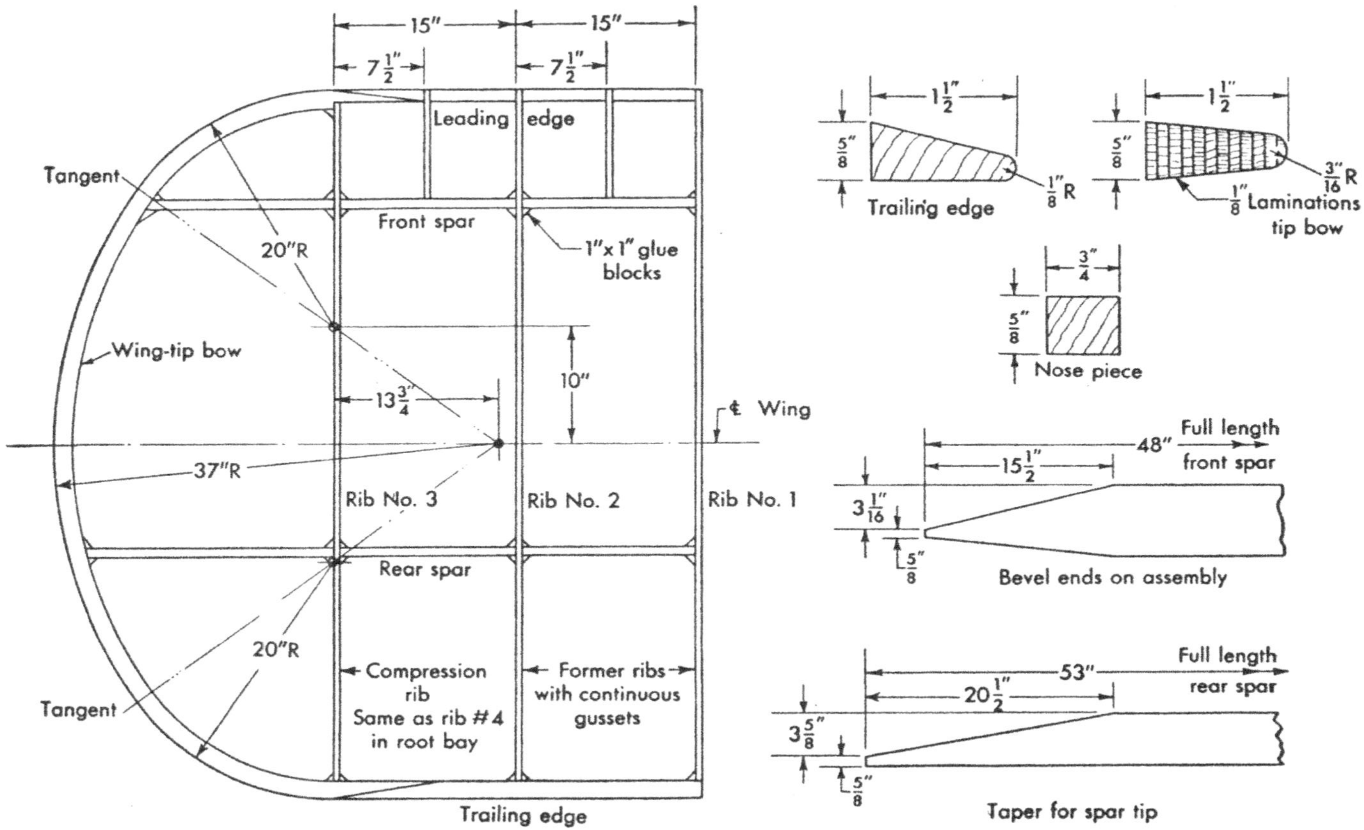

Fig. 150. Plan and detail drawings for the construction of a plywood-covered wing tip.

an inside radius of 37 in. When the ash strips have been prepared, the proper amount of casein glue should be mixed and spread on one side of the first strip and on both sides of each of the other strips except the last strip. The strips should be placed together as soon as the glue is spread and immediately placed in the jig as follows. The straight part of the front end of the bow is clamped into place between the two

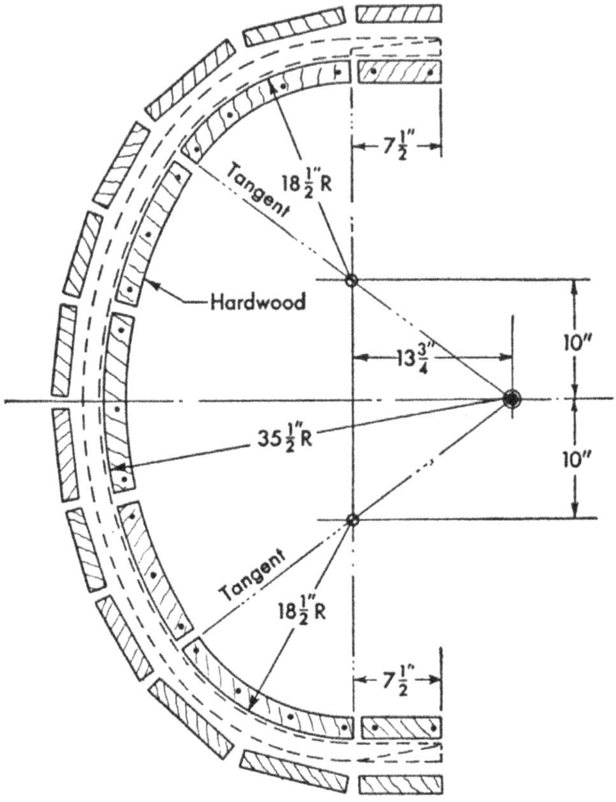

Fig. 151. Construction of a jig for laminating a wing-tip bow.

straight blocks. All of the strips are then bent around the 20-in. radius for approximately one half of the distance when the first block is clamped into place. This procedure is followed all along the bow, clamping each block in place in the proper order. The bow should be allowed to dry in the jig not less than 8 hr., and preferably overnight. After being removed from the jig, the bow should be allowed to dry for an additional 24 hr. before being worked to its proper shape as shown in Figure 150.

Problem 3. Construct the Ribs. This portion of the wing will require three ribs, two form ribs, and one compression rib. The compression rib will be the same as rib No. 4 in the root bay constructed in

Chapter XV, Problem 7. The form ribs will be similar to the ribs constructed in Chapter XV, Problem 4, with the exception that they will be reinforced with plywood gussets extending throughout their entire length. These gussets give the added strength necessary to support the plywood covering. Figure 152 shows the method of reinforcing these ribs by a continuous gusset. Two false ribs will be needed which are

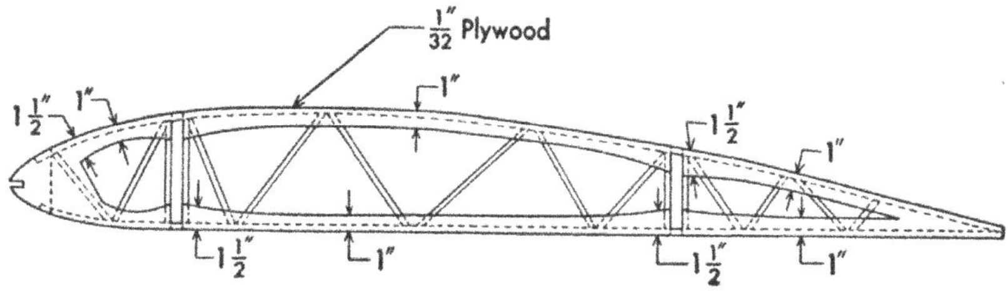

Fig. 152. A continuous gusset for ribs No. 1 and No. 2 in the wing tip.

identical to the false ribs constructed in Chapter XV, Problem 8. In this problem it will not be necessary to rig drag and antidrag wires or drill the holes for fittings. The leading-edge strip and trailing-edge strip as shown in Figure 150 should next be constructed. The glue blocks shown in Figure 150 should be prepared, and the entire bay assembled and glued. After drying not less than 24 hr., install the plywood covering.

Problem 4. Cover a Wing-Tip Bay with Plywood. The covering should be of 3/64 in. mahogany- or birch-faced plywood. It is not practical to cover the entire tip with a single piece of plywood. No horizontal joints are allowed on the leading edge of a wing. A piece of plywood of sufficient size to cover the leading edge between the compression rib and the end of the bay should be formed. This piece will be approximately 30 in. long, and the grain should run around the leading edge from the top of the front spar to the bottom of the rear spar. The edges which will be joined with the other sheets of plywood should be scarfed before forming. When forming, the scarfed edges should be on the inside, except at the outer end, where the scarf should be on the outside. This plywood will be glued by means of a scarf joint to the plywood covering the bay from the front spar to the trailing edge and should overlap. A piece of plywood approximately 30 in. wide of sufficient length to reach from the front of the front spar on the top of the wing to the trailing edge is scarfed on the front and the outer edge. A similar piece is prepared for the bottom of the tip and scarfed in the same manner. A piece of plywood of sufficient size to cover the tip of the wing from the compression rib to the tip is next cut to shape and scarfed on the bottom side where it will join the leading-edge covering and on the bottom side where it joins the covering on the rest of the wing. A similar piece is prepared for the bottom of the wing and scarfed in the same manner. The finish of the surface of the plywood where it will come in contact with the ribs, spars, and wing-tip bow should be scraped or sanded to remove the finish. The pieces covering the portion of the wing back of

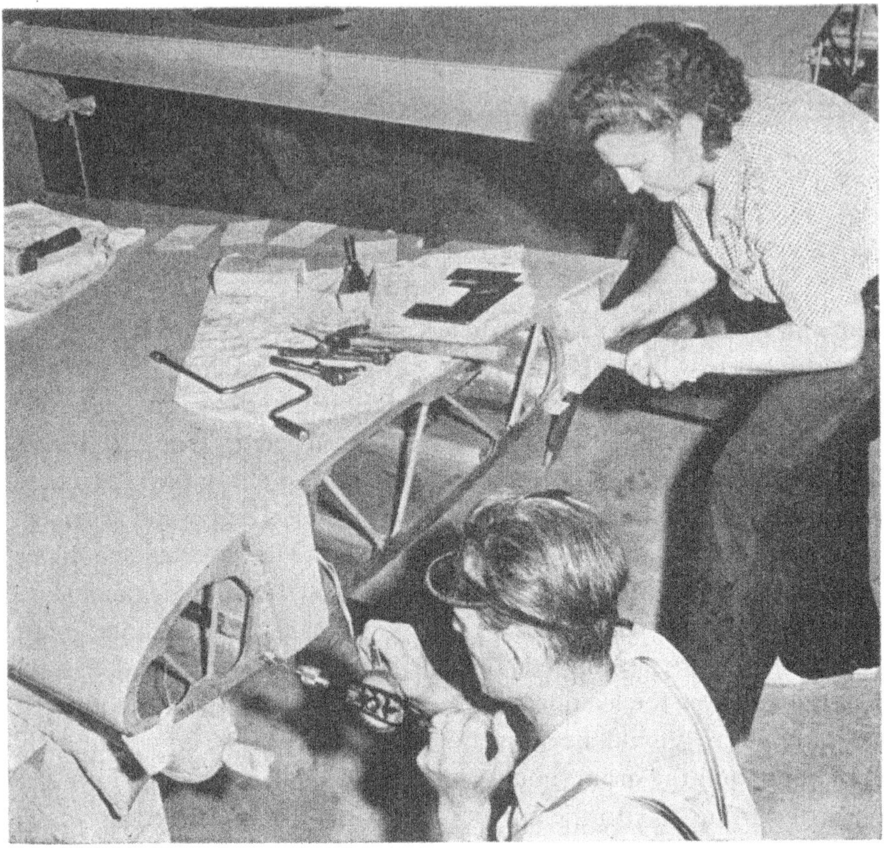

Fig. 153. Outer panel attachment fitting installation on a plywood molded wing. (Courtesy Timm Aircraft Corporation)

the front spar toward the root should now be applied, having first had glue spread on that portion of the internal structure, including the ribs, to which it is to be fastened. It is now strip-nailed to rib No. 1, placing the nails approximately 1 in. apart. Next, nail along the rear spar and the top of rib No. 2; then along the front spar. The edge of the scarf portion should reach approximately to the front edge of the block on top of the front spar between the ribs. This edge should be strip-nailed. The rear edge should next be strip-nailed and then the outer edge strip-nailed with a scarf joint reaching the outer edge of the compression rib. The plywood covering on the bottom of the wing opposite this piece should next be applied in the same manner, and these pieces allowed to dry thoroughly. Remove the nailing strips from the front and outside edge of the pieces already in place and apply the leading edge which has been preformed and scarfed on its edges on the inside and on its outer end on the outside. Strip-nail along the leading-edge strip, ribs, and top and bottom of the front spar. After the leading edge

Fig. 154. A router being used on the molded plywood skin of a wing. (Courtesy Timm Aircraft Corporation)

has thoroughly dried, apply the pieces of plywood to the tip of the wing. This portion of the wing may be covered with a single piece of plywood, top and bottom, or may be covered with separate pieces, dividing the top and bottom into two or three sections. This piece has the inner edge

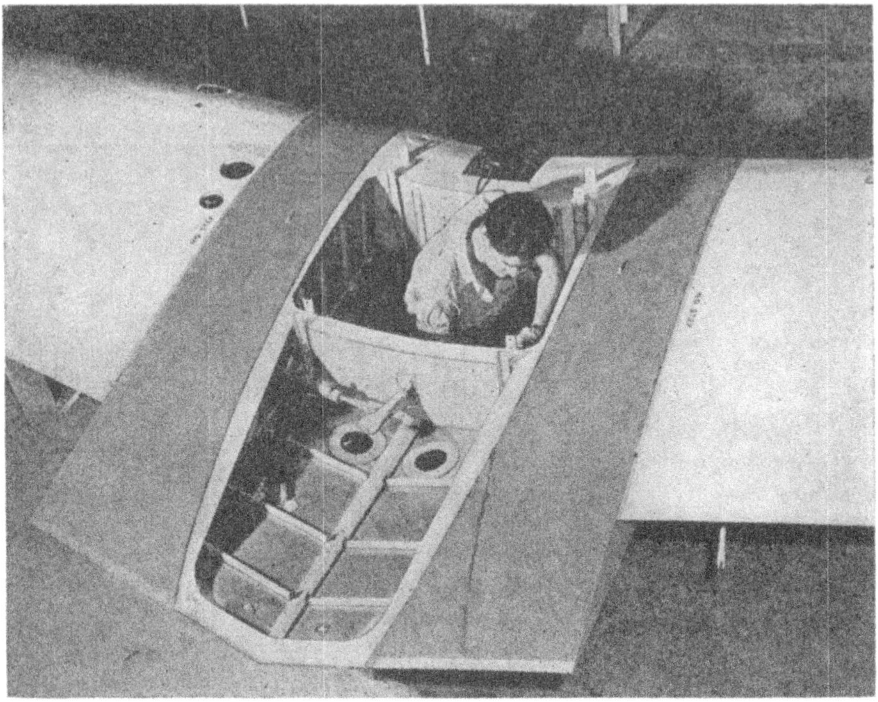

Fig. 155. The center section of a molded plywood wing. (Courtesy Timm Aircraft Corporation)

scarfed on the bottom and is nail-glued over the scarfed edge of the leading edge and the plywood already applied. As soon as it is strip-nailed to the compression rib, the plywood is bent down to the center of the bow by means of hand clamps. These clamps are then placed at the end of each spar and at the center of the 20-in. curved radius, front and rear, and at the front and rear of the compression rib. Nailing strips are placed between the clamps and nails driven at the end and the middle of each strip, entirely around the wing-tip bow. Nails are then driven between the nails already in place until the interval between nails is approximately 1 in. This method of nailing prevents large wrinkles occurring at any point along the wing tip. If sufficient help is available, all of the plywood pieces may be applied and nail-glued in one operation. No part of the structure should remain unnailed more than 20 min. after the glue is spread, unless clamped firmly in place.

XVII CONSTRUCTION OF A FABRIC-COVERED AILERON

It will not be necessary to construct an entire aileron. A portion of an aileron 28⅝ in. long will be constructed and finished on each end as though it were a complete aileron. The ribs in the aileron will not run directly from front to back, but will be inserted, forming triangular trusses as shown in Figure 156. The aileron spar is of clear spruce, ⅜ in. thick, 3¾ in. on the front face, and $3\frac{21}{32}$ in. on the rear face. Two jigs should be constructed to form the end ribs and the angle ribs. The two end ribs are reinforced with $\frac{3}{64}$-in. plywood gussets as shown in Figure 157. One rib will have the gussets on one side, and the other rib will have them on the opposite side. The angle ribs will consist of two righthand ribs and two lefthand ribs as shown in Figure 158.

Problem 1. Construct an Aileron Spar. The aileron spar will be constructed with the dimensions as shown in Figure 157.

Problem 2. Construct the End Ribs. The lefthand end rib should be constructed and the plywood reinforcing nailed in place while the rib is still in the form. After removing from the jig, it is turned over on a flat surface and two small gussets nailed in place. The righthand rib should have the small gussets nailed in place while the rib is in the form, and, after removing from the form, the reinforcing gusset is nailed and glued into place on the opposite side.

Problem 3. Construct the Angle Ribs. The reinforcing gussets on the angle rib are not the same on both sides. The two righthand and two lefthand angle ribs should be constructed. The angle ribs are 12¼ in. long, and the ends are beveled, as shown in Figure 158.

Problem 4. Construct the Trailing Edge. The trailing edge is constructed as shown in Figure 157.

Problem 5. Assemble the Aileron. The aileron is assembled by nailing and gluing the ribs to the aileron spar. Then the trailing edge is glued

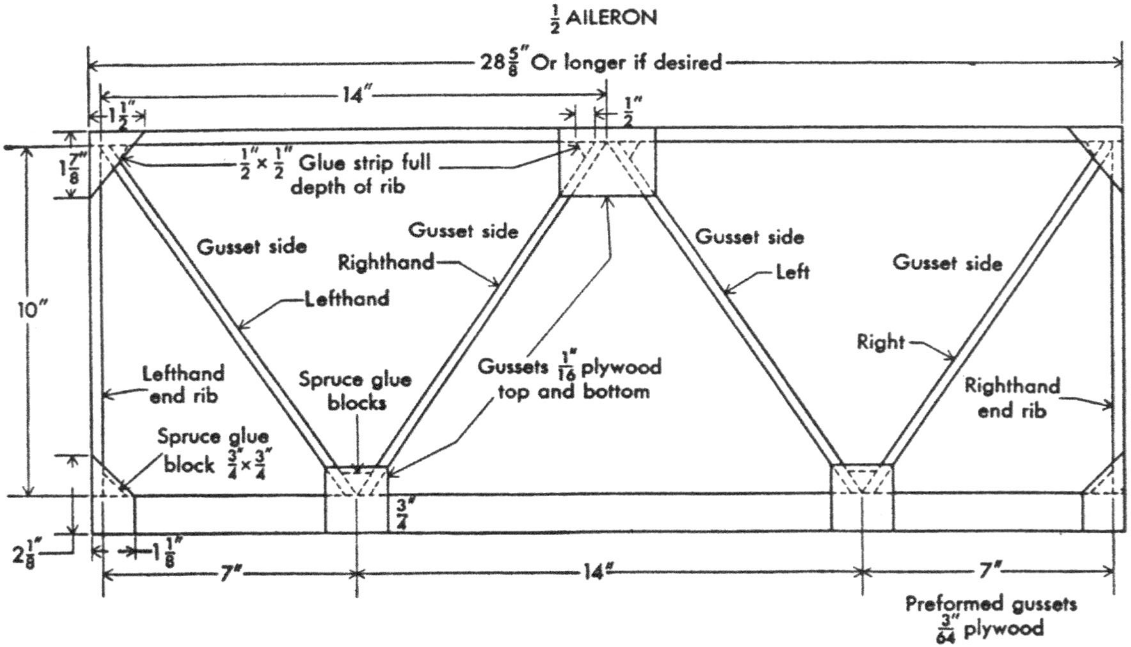

Fig. 156. A plan drawing showing the aileron structure.

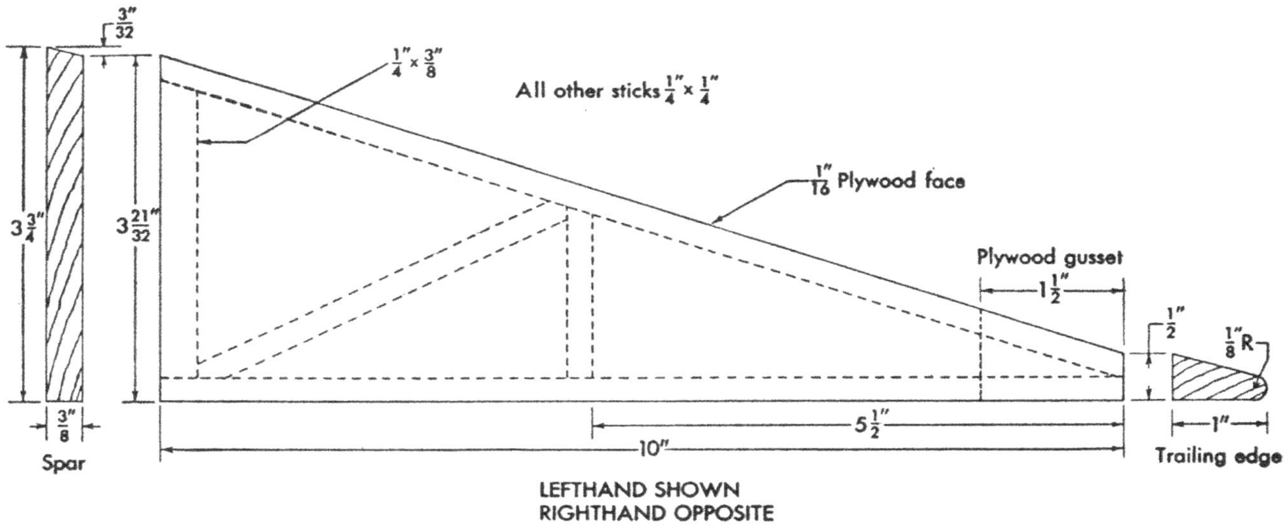

Fig. 157. The construction of the end rib of an aileron.

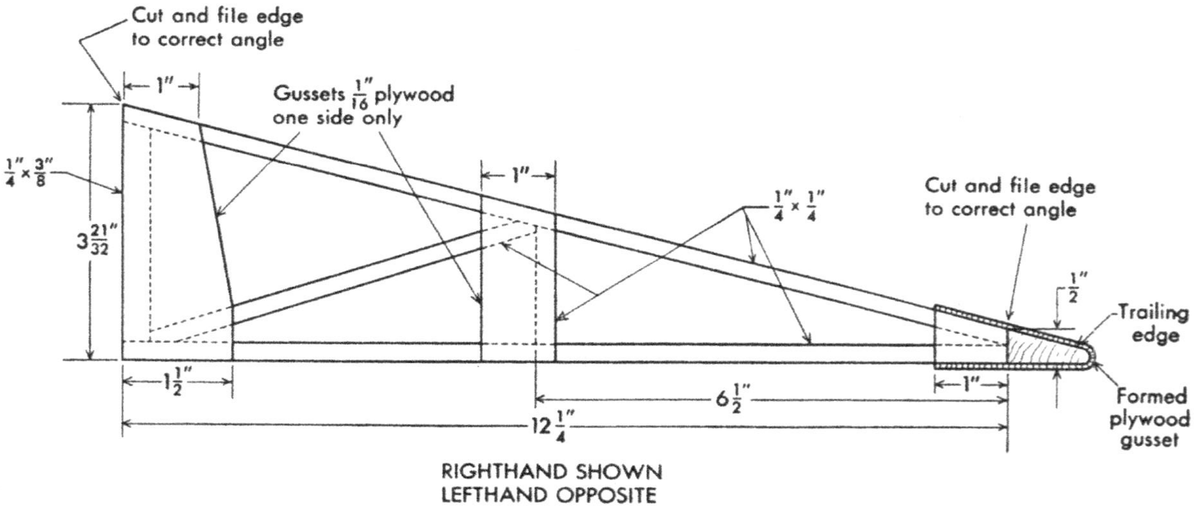

Fig. 158. The construction of an aileron angle rib.

and nailed into place, fastening with 3/64-in. plywood gussets which have been preformed, extending from the root of the rib on top, around the trailing edge to the end of the rib on the bottom, as shown in Figure 156. The aileron is then covered with fabric, as described in Chapter XV, Problem 14, and finished in the same manner as the fabric on the root bay.

XVIII CONSTRUCTION OF A PLYWOOD-COVERED VERTICAL FIN

The vertical fin aids in the directional stability of the airplane. It is usually a fixed surface and not adjustable in flight. In most light airplanes the vertical fin is not adjustable and is fixed solidly to the fuselage structure. The vertical fin constructed in the following problems is of this type. This fin is of all-wood construction and covered with plywood. The parts to be constructed are the rear vertical member, called the main spar, which is fastened at its lower end to the tailpost of the fuselage; the top and leading edge, which is formed of laminated spruce strips; four horizontal ribs; vertical bracing strips; and glue blocks.

Problem 1. Construct the Main Spar. This member is to be made from clear, straight-grained spruce having a total over-all length of 73¾ in. It is ⅜ in. thick and 4⅜ in. wide at its widest part. Select a piece of spruce having not less than 6 annual rings per inch. This piece should be edge grain, 4½ in. wide, ½ in. thick, and approximately 78 in. long. Dress one side and one edge until they are smooth and straight and at right angles to each other. Square one end and dress the piece down until it is exactly 4⅜ in. wide and ⅜ in. thick. Mark off on the square end a distance on each edge of exactly 21$\frac{11}{32}$ in. Draw a line across the face exactly 73¾ in. from the squared end. Find the center of this line. On each side of the center make a mark $\frac{9}{16}$ in. from the center. Connect this point with the line on the edge 21$\frac{11}{32}$ in. from the squared end. Do the same with the other mark. These two lines will taper the vertical member from a point 21$\frac{11}{32}$ in. from the squared end to 1⅛ in. wide at a point 73¾ in. from the squared end. With a plane, dress down each side of the main spar to these lines, leaving the excess length at the small end to assist in keeping the end true. When the tapered part is down to the dimensions shown, cut to exactly 73¾ in., by sawing off the excess material at the tapered end, and the main spar is complete.

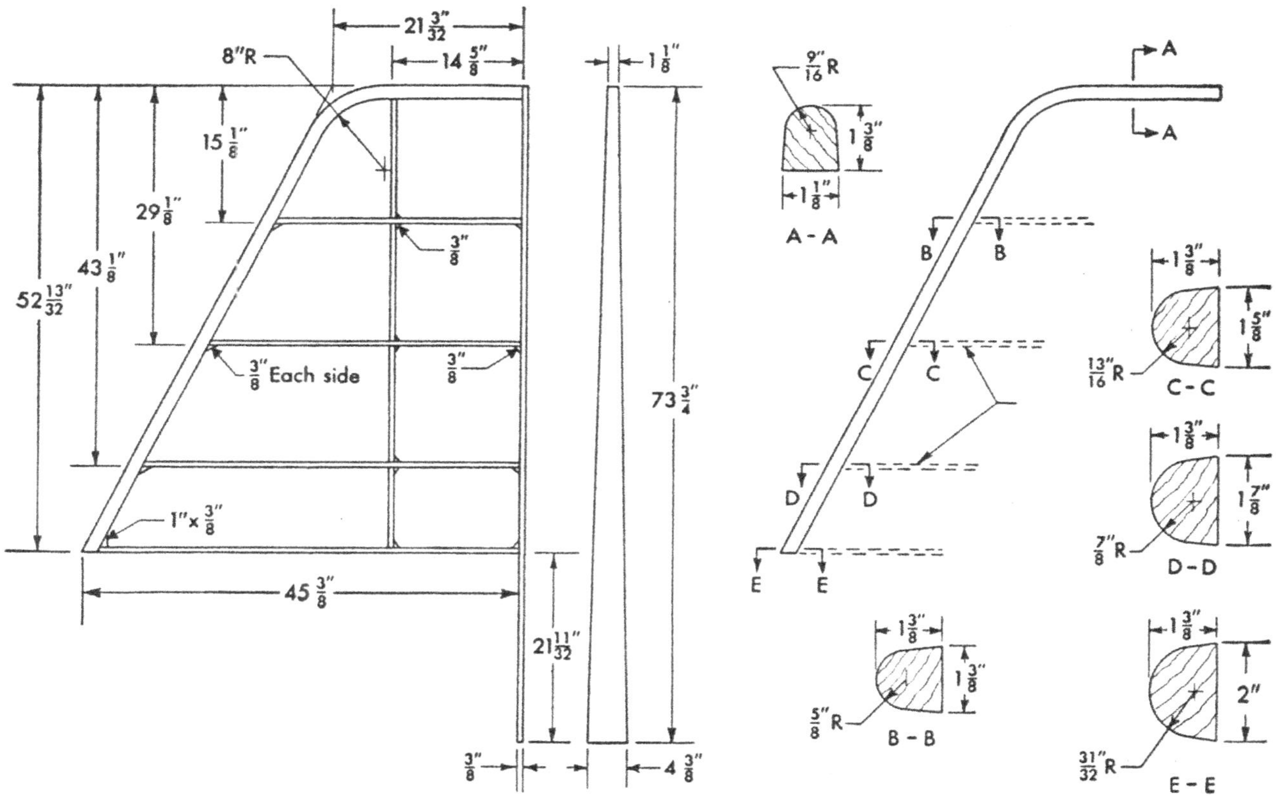

Fig. 159. The construction of a plywood-covered vertical fin.

Problem 2. Construct the Laminated Top and Leading Edge. A jig must be prepared to laminate the top and leading edge. This jig is similar to the jig used in forming the wing-tip bow. This part of the fin will be formed of 1/8-in. spruce laminations 2 in. wide. On a bench top or other smooth surface sufficiently large, lay out the inner edge of the laminated member. You will note from Figure 159 that this member is straight for a distance of 14 5/8 in., then curves on an 8-in. radius to a straight portion extending to a point 45 3/8 in. from the main spar. Two-inch blocks should be fastened firmly to the jig base, as shown in Figure 160. The laminations should be of straight-grained spruce, vertical-grained on the wide face. Eleven laminations 80 in. long, 1/8 in. thick, and 2 1/16 in. wide will be needed. The extra 1/16 in. should be allowed for smoothing and dressing after removing from the form. A sufficient number of C-clamps and backing blocks should be on hand before starting the gluing operation. The blocks which hold the curved part should be shaped to the finished curve of the member as shown in Figure 160. Apply casein glue evenly to both sides of each strip with the exception of the two outside strips, piling them one on top of the other as the glue is spread. Beginning at the top, clamp the part that is 14 3/8 in. long into place, screwing the clamps up firmly. Carefully bend the remaining strips partly around the curved block, and place the first curved block in position and clamp. Bend still further around the curve, and place the second curved block into position and clamp. Be sure that the edges of all of the strips are even. Finish bending around the curve, clamping the curved blocks into place. Now apply the long, straight block, tightening all of the clamps firmly. After about 20 min., go over all the clamps and retighten to make sure that sufficient pressure is applied. The pressure applied should be approximately 150 lb. per sq. in. Allow the member to dry not less than 8 hr. before removing

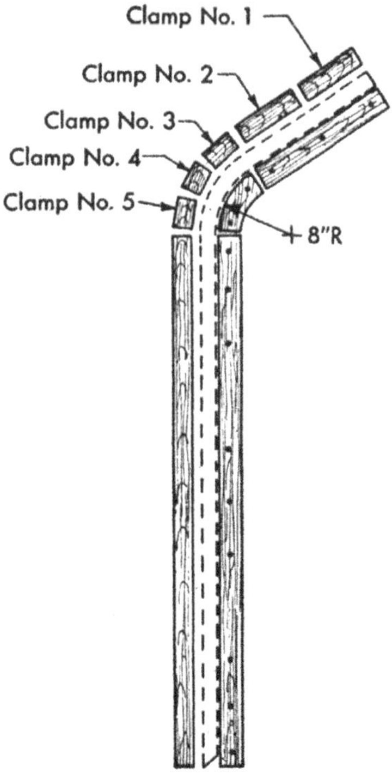

Fig. 160. The construction of a jig for a vertical-fin leading edge.

from the form. It is better to leave the member in the form 24 hr., if possible. Before dressing down to final form, the member should be allowed to dry at room temperature for 2 or 3 days to allow even distribution of moisture. The fabricated member should then be dressed down to the form shown on the cross-sectional drawings, Figure 159. Note that the edges are tapered from the bottom which is 2 in. wide, to 1⅜ in. wide at the top rear end. The bottom front end is smoothed off with a radius of 3⅓₂ in., and the top rear end to a radius of ⅝ in. The thickness remains at 1⅜ in. throughout.

Problem 3. Construct the Fin Ribs. There are four ribs to be constructed. The top rib is to be constructed of spruce cap strips 5/16 in. thick and ⅜ in. wide, edge grain on the larger dimension. The top rib has an over-all length of 26¾ in. It is 2 in. wide at the rear and outside dimension, and is 1⅜ in. wide at the front-end, outside dimension. A jig will not be necessary to construct this rib. This rib is to be covered on the top side with plywood 3/64 in. thick. This plywood may be of any plywood available in this thickness, but mahogany outside layers are preferred. The plywood is cut the shape of the rib as shown in Figure 161. The end pieces of ⅜-in. by 5/16-in. spruce are cut to the proper length. A cross piece should be placed 12⅜ in. from the small end of the rib as shown in Figure 161. The cap strip itself can now be bent into place and fastened at each end with two plywood gussets, as shown in Figure 161. The various cross pieces should be fastened in the same way with gussets of 3/64-in. plywood glued and nailed. The rib is then turned over, the top surface is coated with casein glue, and the plywood which has been cut to shape is nail-glued to the rib as shown in Figure 161. The other three ribs are constructed in a similar manner.

Problem 4. Construct the Vertical Bracing Members. There are two vertical bracing members which extend from the bottom rib to the top member of the fin. These strips are of ⅜-in. by 5/16-in. spruce and are fastened in place by the use of saddle gussets and glue blocks, as shown in Figure 161.

Problem 5. Prepare the Glue Blocks. The following glue blocks are required: 5 for fastening the ribs to the main spar at the end; 16 to fasten the vertical bracing members in place; and 4 to fasten the ribs to the leading edge. All of these blocks are prepared according to the dimensions given in Figure 159.

Problem 6. Assemble the Vertical-Fin Framework. The vertical fin is assembled by nailing and gluing the four ribs to the main spar as shown

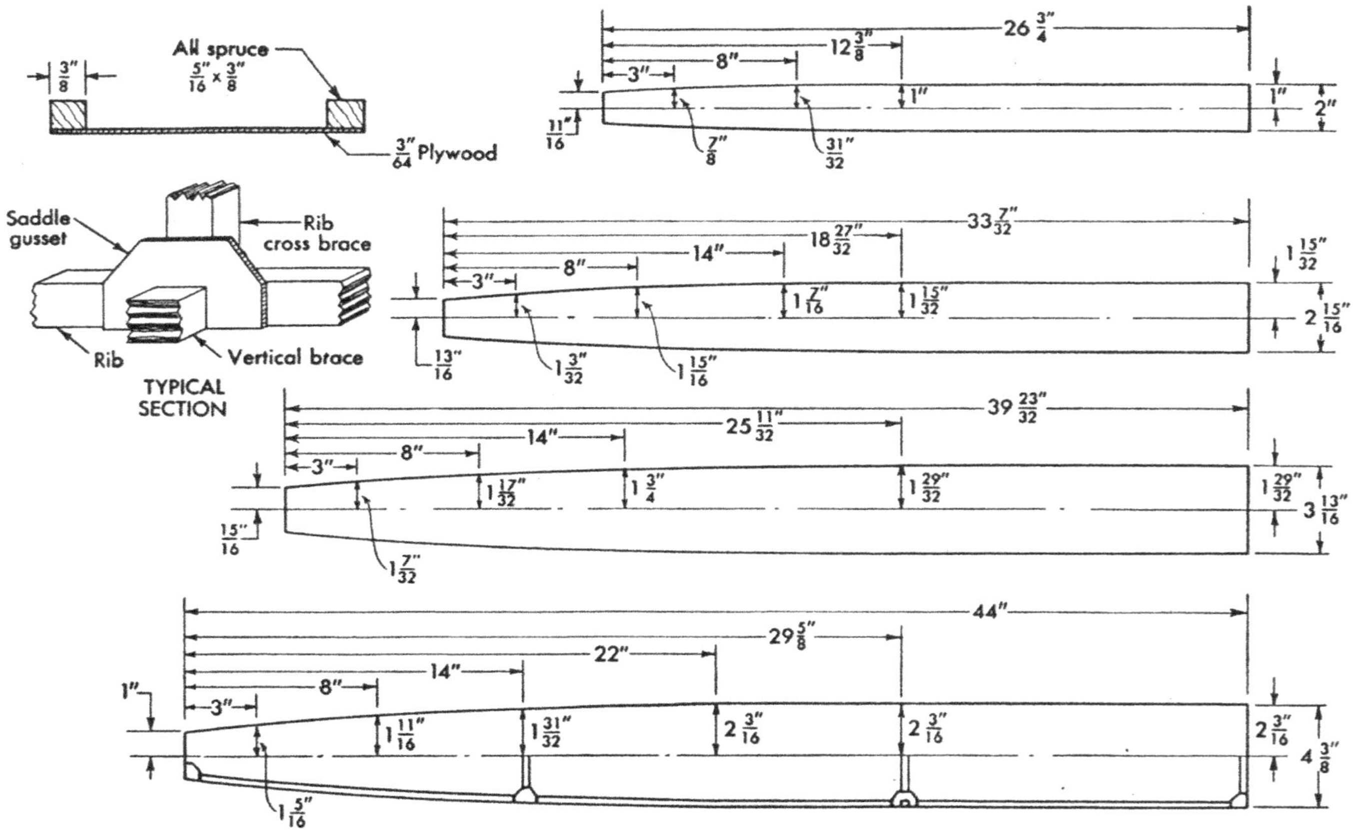

Fig. 161. The construction of the ribs for a vertical fin.

in Figure 159. The top of the rear member is then glued and nailed to the laminated edge, followed by the front end of the ribs. With the assembly firmly clamped in place, insert and fasten the vertical braces as shown in Figure 159. Allow to dry not less than 8 hr.

Problem 7. Apply the Plywood Covering. The whole fin is to be covered on both sides with plywood $\frac{3}{64}$ in. thick, preferably mahogany outside veneer, although any plywood of this thickness available may be used. The plywood should extend from the rear edge of the main spar, and from the bottom edge of the bottom rib, and be of such size as to cover the leading and top edge to the curved portion. The grain on the outside veneer should be placed vertically. The edge of the ribs, the main spar, and the leading and top edge should be coated with casein glue, and the plywood covering clamped along the leading and top edge with C-clamps, as shown in Figure 162. The plywood covering is then nailed in place, as shown in Figure 162, and allowed to dry not less than 24 hr. After thoroughly drying, the edge of the plywood along the front and top edge should be dressed down to make a smooth, tapered joint with the leading and top edge. This can be done with a fine wood rasp and sandpaper.

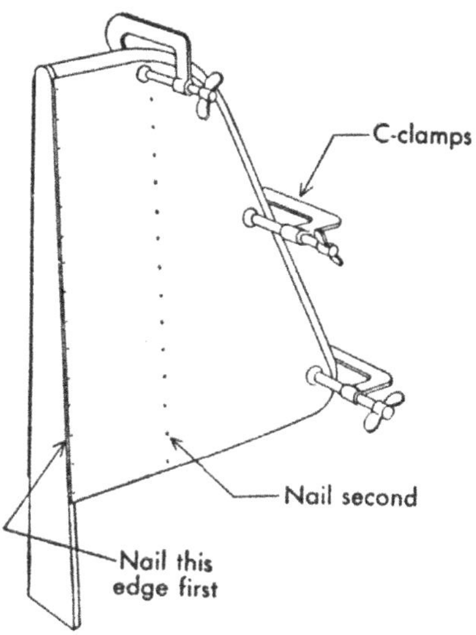

Fig. 162. Applying the plywood covering to a vertical fin.

XIX CONSTRUCTION OF A FABRIC-COVERED RUDDER

The control surfaces are usually of lighter construction than the stabilizing surfaces. The construction of this rudder is typical of those used even when the rest of the airplane is covered with plywood.

Problem 1. Construct a Rudder Post. The rudder post is made of clear spruce 37½ in. long, 4 in. wide at one end, ⅝ in. wide at the other end, and ⅜ in. thick, as shown in Figure 164.

Problem 2. Construct the Trailing Edge. The top and trailing edge of the rudder is laminated from clear white ash ⅛ in. thick and ¾ in. wide. Twelve strips will be needed. At the top edge, this lamination is straight for 6¼ in., then curves with an outside radius of 7½ in. through approximately 75° and then is straight to the point of junction with the bottom

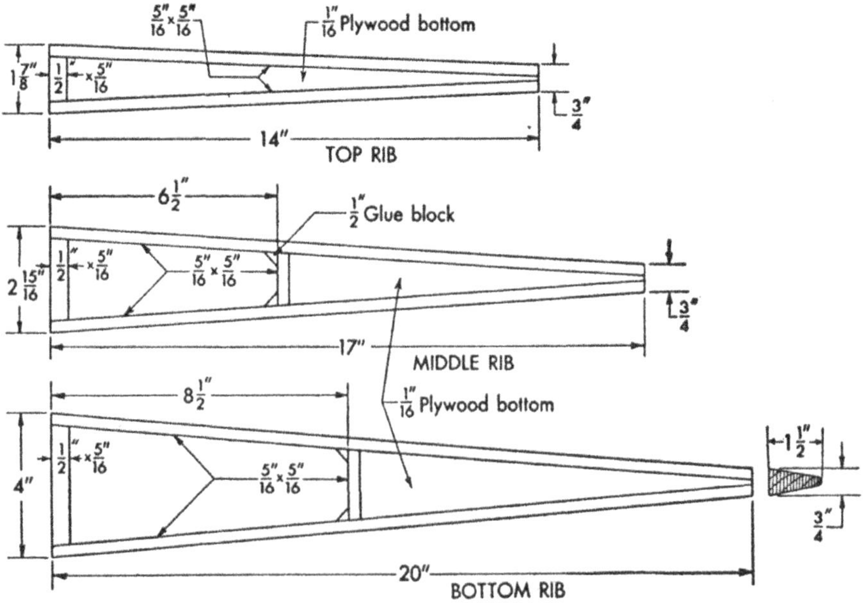

Fig. 163. The rib construction for a rudder.

rib of the rudder. A jig will be necessary, and is built similar to the jig used in Chapter XVIII, Problem 2. After the trailing edge has dried not less than 24 hr., it should be shaped as shown in Figure 164.

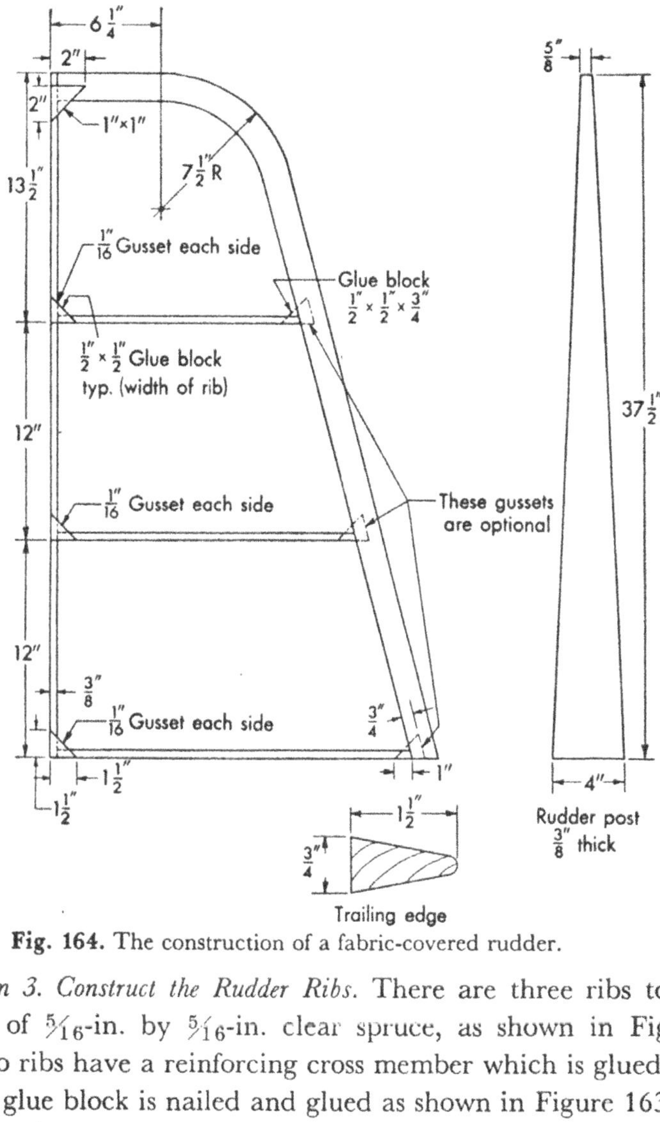

Fig. 164. The construction of a fabric-covered rudder.

Problem 3. Construct the Rudder Ribs. There are three ribs to be constructed of 5⁄16-in. by 5⁄16-in. clear spruce, as shown in Figure 163. Only two ribs have a reinforcing cross member which is glued in place. A ½-in. glue block is nailed and glued as shown in Figure 163.

Problem 4. Assemble the Rudder. The rudder is assembled by nailing and gluing the ribs to the rudder post, then the leading edge is glued and nailed to the other end of the ribs, as shown in Figure 164. The glue blocks are glued and nailed into place, and gussets applied at each end of the ribs, as shown in Figure 164. After drying 24 hr., the rudder is covered with fabric and finished as described in Chapter XIII.

XX CONSTRUCTION OF A PLYWOOD-COVERED HORIZONTAL STABILIZER

It will not be necessary to construct both halves of the stabilizer. This problem will be to construct the left half of the horizontal stabilizer which is to be covered with 3/32-in. plywood having a surface veneer of mahogany or birch. The leading edge and the end will be formed of laminated white ash. These laminations are 2¾ in. wide, ⅛ in. thick, and 74 in. long. Twelve strips will be needed. A jig will be necessary, similar to the jig used to form the laminated top and leading edge of the vertical fin in Chapter XVIII. (See Figure 160.)

Problem 1. Construct the Laminated Leading Edge. The laminations are prepared with the grain vertical to the wide face. The method of fabrication is the same as that used for the construction of the laminated edge of the vertical fin, except that the gluing pressure should be somewhat greater — from 175 to 200 lb. per sq. in. The leading edge is straight for 39 in., then bent on an 8-in. inside radius to approximately 105°, and then straight to where it joins the main spar of the stabilizer. The straight portion at the outer end is first clamped in place, then the portion forming the curve, and finally the straight portion of the stabilizer leading edge. Beginning at the point 39 in. from the fuselage, the leading edge is tapered evenly down to 1 in. wide and 1 in. thick at the point where it meets the spar.

Problem 2. Construct the Spar. The spar, which is the rear member of the stabilizer, is made of ½-in. spruce, edge grain on its wide surface. This piece is finished in the same manner as the main spar of the vertical fin in Chapter XVIII. Figure 165 shows the method of construction and the size of materials used. Two strips extend from the fuselage end to the tip of the stabilizer and are constructed and installed in a similar manner to the braces in the fin constructed in Chapter XVIII. Figure 165 shows the method of fastening and the gluing blocks needed.

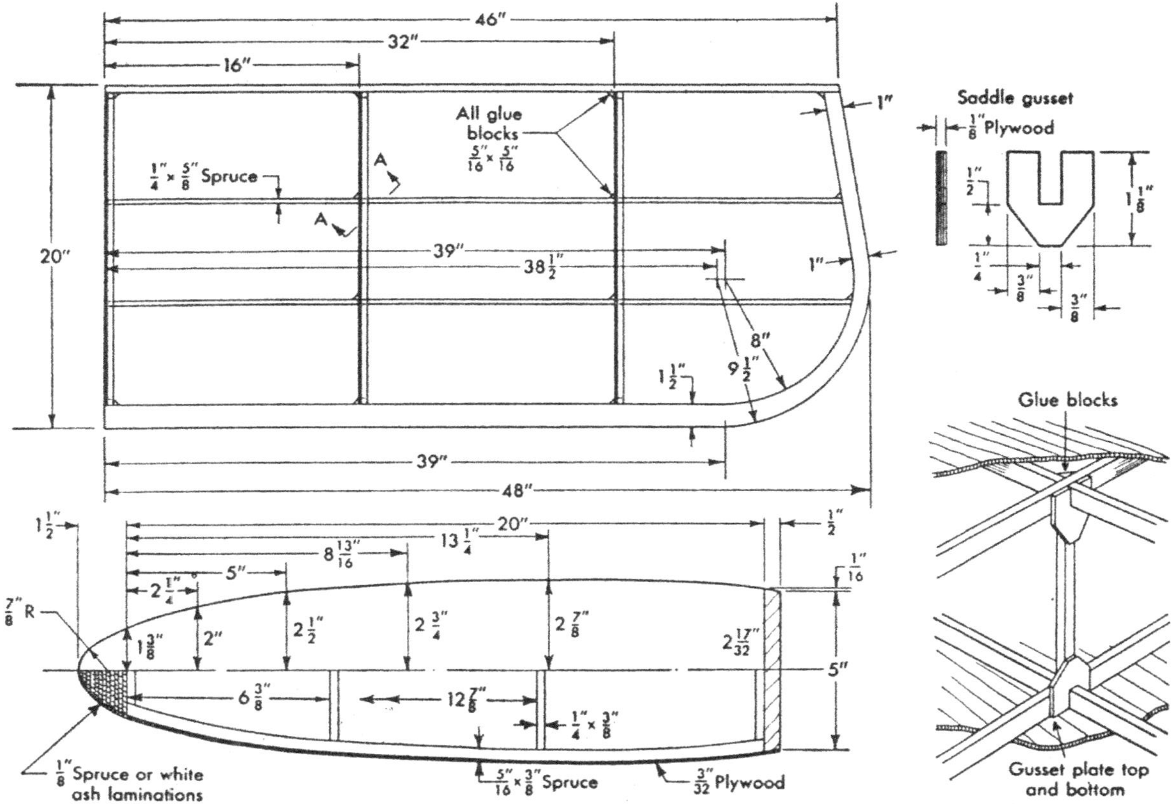

Fig. 165. The construction of a plywood-covered stabilizer.

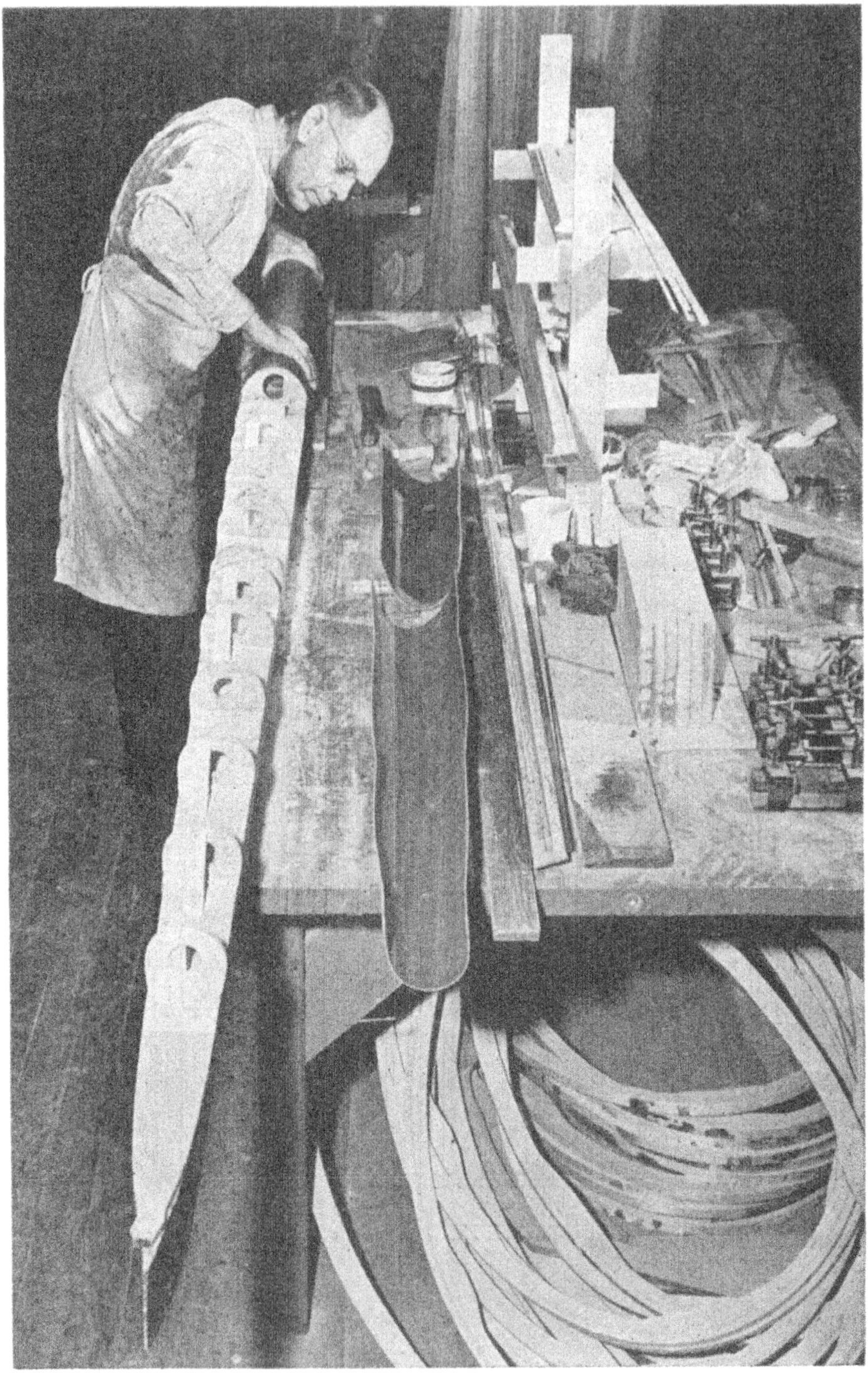

Fig. 166. Applying a plywood leading edge. (Courtesy Timm Aircraft Corporation)

CONSTRUCTION OF A PLYWOOD-COVERED STABILIZER

Problem 3. Cover the Stabilizer. The stabilizer is to be covered with 3/64-in. mahogany-faced plywood. The plywood may be applied in one piece by being bent around the leading edge or preformed by using a suitable form. The straight portion from the fuselage to the beginning of the curved portion should first be glued and nailed in place, then nailed along the ribs and rear spar. The tip portion is then glued to the laminated member and held in place by hand clamps while being nailed. The clamps should remain in place until the glue has thoroughly set, not less than 6 or 8 hr.

XXI CONSTRUCTION OF THE REAR SECTION OF A PLYWOOD–COVERED FUSELAGE

By using plywood as stressed-skin covering it is possible to construct a fuselage entirely of wood. A fuselage having no diagonal cross members is called "monocoque construction." The construction of all bays of a fuselage of this type is similar throughout. The building of a part of a fuselage consisting of three rear bulkheads will familiarize the workman with this type of construction. The bulkheads should be laminated from 1⁄32-in. mahogany-faced plywood, and each bulkhead should have five laminations, making a total thickness of 5⁄32 in. The eight longerons are of clear hickory. Ash or Douglas fir may be substituted if hickory is not available. This section is to be covered with 3⁄64-in. plywood, either birch- or mahogany-faced.

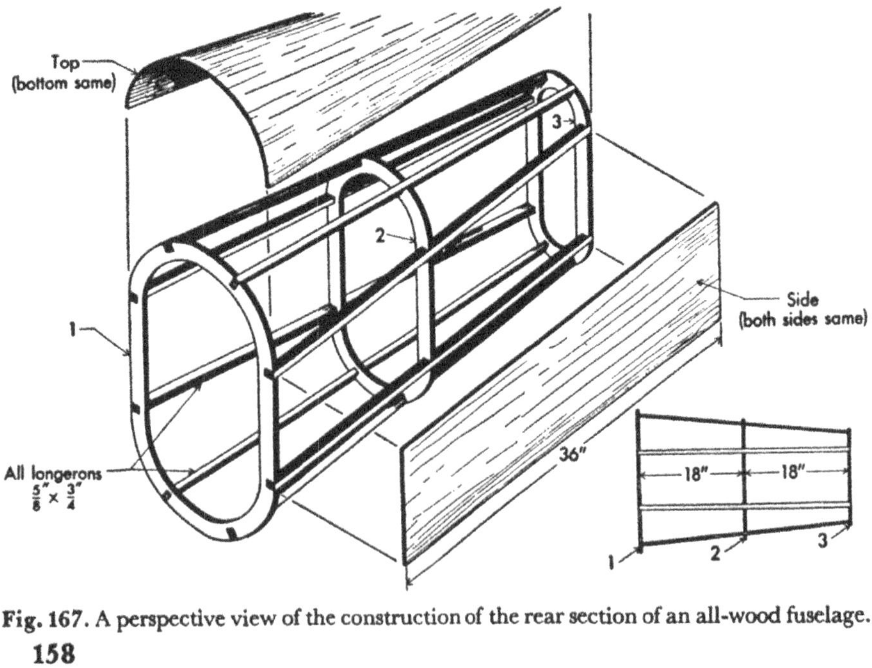

Fig. 167. A perspective view of the construction of the rear section of an all-wood fuselage.

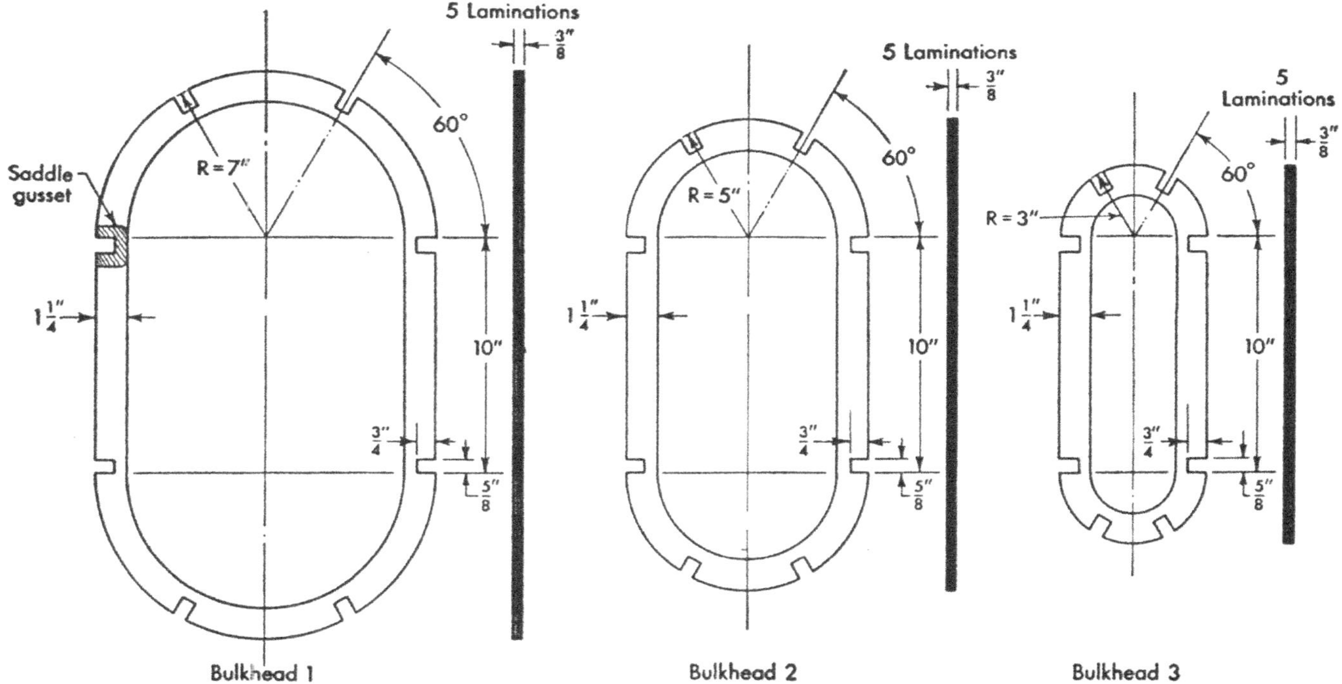

Fig. 168. The construction of bulkheads for a fuselage section.

Problem 1. Construct the Longerons. The eight longerons are constructed of clear hickory, rectangular in shape, ⅝ in. thick, ¾ in. wide, four being 36 in. long, and four being slightly longer due to the slope of the fuselage. These members should have edge grain on the wide face.

Problem 2. Construct the Plywood Bulkheads. It will not be necessary to build a jig to construct the bulkheads. Each layer of the bulkhead may

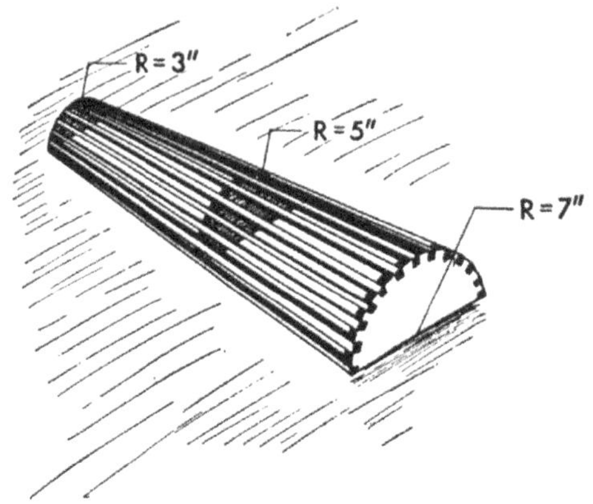

Fig. 169. A form for preforming the plywood covering for the top and bottom of a fuselage section.

be built up of several pieces of plywood joined together with a scarf joint which should have a slope of 1 in 20. The joints should be so arranged that the joints in the various layers do not fall in the same places in the lamination. It would be possible to saw all laminations in a single piece, but this would entail a great waste of plywood. The sides of the plywood which are to be glued together should be lightly scraped or sandpapered to remove the glazed finish. The proper amount of casein glue should be spread on each face of the glued layers, and pressure applied to the entire bulkhead. Approximately 175 lb. pressure per square inch should be applied. Allow 8 hr. for drying; 24 hr. additional drying should be allowed before notches for the longerons are cut. Reinforcing saddle gussets should be prepared as shown in Figure 165. Sixteen of these gussets will be necessary for each bulkhead.

Problem 3. Assemble the Fuselage. The structural part of the fuselage is assembled by gluing the longerons into the notches and clamping into place. The saddle gussets should be placed on each side of the bulkhead and nailed and glued as soon as the longerons are in place.

Problem 4. Apply the Plywood Covering. The plywood covering, which consists of $\frac{3}{64}$-in. mahogany-faced plywood, should be applied in four pieces: the two pieces covering the sides of the fuselage are flat and rectangular as shown in Figure 167; the other two pieces should be preformed as shown in Figure 167. The sides of the pieces should be scarfed on the top and bottom edges and glued and nailed into place. The preformed plywood is then applied in a similar manner, forming a scarf joint with the side pieces. Care should be taken to obtain a good scarf joint. The plywood skin may be finished as described in Chapter XIII.

Figure 169 shows a form which may be built to shape the top and bottom pieces of plywood.

XXII WOOD PROPELLERS

The balance and alignment requirements for propellers are probably the most exacting requirements in the manufacture of aircraft. Balance and freedom from warping in service depend as much upon matching all laminations for density, moisture content, and direction of annual growth rings as upon the species of wood used. Some species give very satisfactory results with rapid manufacture and only a moderate degree of refinement, while others, such as oak, must be carefully handled, with ample time for conditioning, to give the best service.

Very few wood propellers are made from a single piece of wood. Most are built up by gluing together several pieces of wood, making a laminated structure. The glue used may be resin or casein glue. The moisture content of the wood at the time of gluing should be between 7 and 9 per cent. All laminations should have as nearly as possible the same moisture content. All laminations in a single propeller should be of either flat-sawed or quarter-sawed material. Flat-sawed and quarter-sawed lumber should not be used in the same propeller. The adjacent laminations are arranged so that the growth rings in one are reversed in position and direction to the other. This tends to minimize distortion and stresses. The laminations may be cut to such shape that, when combined, they will form the approximate shape of the propeller, or the dimensions may be equivalent to the over-all length and width of the propeller. Making all laminations the same width tends to waste material. Propeller material should be carefully selected, being straight grained and free from any defects which might weaken the wood. Usually a hole is bored at a point where the center of the hub will be located, before gluing. A loosely fitting bolt may be placed through this hole, and the various laminations tested for balance. A shifting of laminations or turning from end to end may assist in getting a preliminary balance for the whole propeller before gluing. A final surfacing of the laminations should be made just before gluing. The surface of each lamination should be smooth and even. Each lamination should be uniform in thickness throughout its length and width. The propeller woods are hardwoods and require more attention to details of gluing than softwoods. Whenever possible, the glue should be spread by machine. If spread by hand, extreme care must be used that the proper amount of glue is applied. Six and one half to 7½ lb. of casein glue per 100 sq. ft., or 4½ to 5 lb. of cold-setting resin glue should be used. The laminations should be placed together as soon as the glue is spread. Pressure should be applied in less than 15 min. after the first glue is spread. The wood should be at a temperature of from 70° to 80° F. for casein glue. A gluing pressure of approximately 200 lb. per sq. in. is recommended. It is difficult to obtain sufficient pressure by the use of hand clamps. Heavy gluing blocks and small jackscrews are most desirable. This method requires a heavy framework. After removal from the press, the propeller should be allowed to condition for a period of approximately 10 days. Propellers are usually shaped by machinery with the final shaping and smoothing done by hand. Templates are used which fit the different cross sections of the propeller to assure the proper shape of the propeller blade.

Propellers must be carefully balanced to prevent vibration. The two blades of the propeller should be equal in weight, and the weight should be distributed as equally as possible in the blades. Usually the tips and leading edges of wooden propellers are protected by a metal covering, and many have a cloth covering over the outer part of the blade. The cloth is woven to the proper shape and cemented into place. Propellers are also made from compressed wood or densified wood and are molded in a single piece. Only minor repairs to wooden propellers should be made in the field. Major repairs may be made only by the manufacturer or at an approved repair station.

XXIII APPROVED REPAIRS

All repairs to certificated aircraft must either be made in a manner approved by the Civil Aeronautics Administration or be inspected by an inspector of the Administration. The Administrator of the Civil Aeronautics Administration has designated who may perform approved repairs: (1) the manufacturer of the aircraft; (2) an approved repair station; or (3) a certificated aircraft mechanic.

There are two classes of aircraft repairs.

Minor Repairs. These consist of repairs to nonstructural members such as cowling, turtle backs, wing and control surface fairings, electrical installations, windshields, and patching of fabric and the replacing of fabric-covered surfaces involving an area not greater than required to cover 2 adjacent ribs. Minor repairs to plywood-stressed covering may be made when ribs or structural members are not directly affected, but should not exceed 3 in. in any direction. Minor repairs also include replacement of complete components or units, when such parts are supplied by the original manufacturer or are manufactured in accordance with approved drawings.

Major Repairs. These are complex repairs of vital importance to the airworthiness of an aircraft and include the replacing, strengthening, reinforcing, and splicing of structural members such as spars or parts of spars, main truss-type beams, the main ribs of wings, compression members, wing or tail surfaces, brace struts, fuselage, longerons, bulkheads, side trusses or horizontal trusses; repairs to 3 or more adjacent ribs, or the leading edge of wings covering 3 or more main ribs, repairs of fabric requiring the covering of more than 2 adjacent ribs, the repair of damaged plywood-stressed covering exceeding 3 inches in any direction, the repair of portions of skin sheets by making additional seams, or the splicing of skin sheets.

In making any repairs, approved workmanship and techniques must be used. All materials used must be of equal or better quality than the original material, and the repair must always equal or exceed the original strength.

Emergency Repairs. It must be understood by the workman that emergency repairs should be replaced by permanent repairs as soon as possible.

Emergency Repairs to Skin. Most emergency or temporary repairs are made to aircraft covering. These repairs are often made to prevent further damage from wind or weather. If the damaged area of plywood is small, fabric may be doped to the plywood covering the hole. This is not usually recommended if the hole exceeds, in its greatest diameter after trimming, 40 times the thickness of the plywood. This method of repair is not recommended if the hole is in the leading edge or front area of the fuselage where it is subjected to high air pressures.

Holes of small size may be covered with temporary plywood patches. The jagged holes should be trimmed out, and the edge of the patch beveled to reduce resistance to the air.

Damage to leading edges and other parts of the structure subjected to high air pressure may be covered with plywood. The holes should be trimmed out to a smooth line and covered with plywood having the proper strength. Whenever possible, the holes should be trimmed out to such an extent that the edges of the plywood patch rest on supporting members in the inner structure. For example, a hole in the leading edge of a wing should be trimmed out to such an extent that the plywood patch would have a bearing surface on two adjacent ribs. Plywood may be soaked if necessary to aid in bending, using the form ribs or adjacent edges of the leading edge as a form. Plywood should be glued and nailed in place. It is not necessary to match grains, as this repair is only made to enable the airplane to be flown to a base where permanent repairs may be made. Cracked or broken ribs may often be temporarily repaired by nailing on pieces of material which act as splints, or by tacking a piece of plywood over one entire side of a broken rib. In an emergency, a wing bow or other similar structure may be replaced by a curved piece of wood, such as a piece sawed out of solid lumber.

Permanent Repairs. All permanent repairs are made in such a manner that they are expected to become a permanent part of the aircraft and should, therefore, be made at least as strong as the original part and finished in an approved manner. All permanent repairs must be made in accordance with the Civil Air Regulations as established by the Civil Aeronautics Administration.

XXIV APPROVED FABRIC REPAIRS

Problem 1. Repair a Straight Tear in the Fabric Covering of the Wing. Remove the finish from the fabric with a dope thinner or some suitable solvent for at least 1 in. in all directions from the damaged portion. The edges are drawn together using a baseball stitch. The thread should be the same as the thread used for sewing the fabric covering and should be locked by using the seine knot. The patch should extend 1 in. beyond the edge of the tear in all directions, and the edges may be pinked, frayed, or plain. The patch should be of the same material as the fabric being repaired. A coat of clear dope should be applied over the repair, and the patch pressed into the dope while it is still wet. The patch is then covered with a coat of clear dope and allowed to dry 30 to 45 min. The repaired area may then be finished in the same way as the original fabric.

Problem 2. Repair a Triangular Tear. Remove the original finish from the area to be covered by the patch with dope thinner or other solvent.

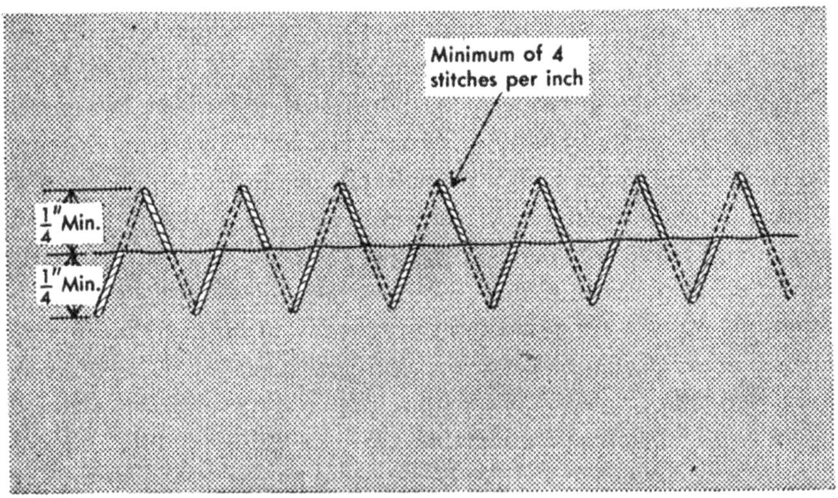

Fig. 170. Baseball stitch used in fabric repairs.

The edges of the tear are then drawn together using a baseball stitch as shown in Figure 170. The patch is then prepared similarly to the patch in Problem 1, extending beyond the repair at least 1 in. on all sides.

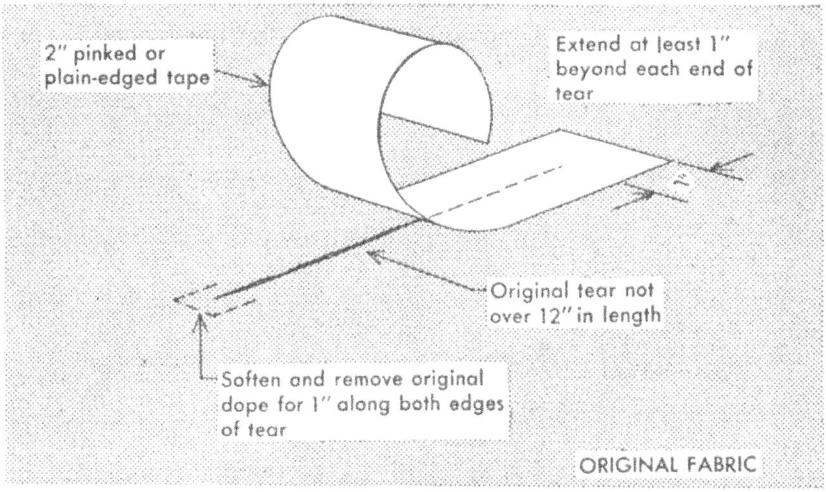

Fig. 171. A fabric patch for small tears not over 12 in. long.

Problem 3. Repair a Hole in a Fabric-Covered Wing. Small holes in a fabric-covered wing having a diameter of less than 3 in. in any direction may be repaired by cutting out any ragged edges. The original finish

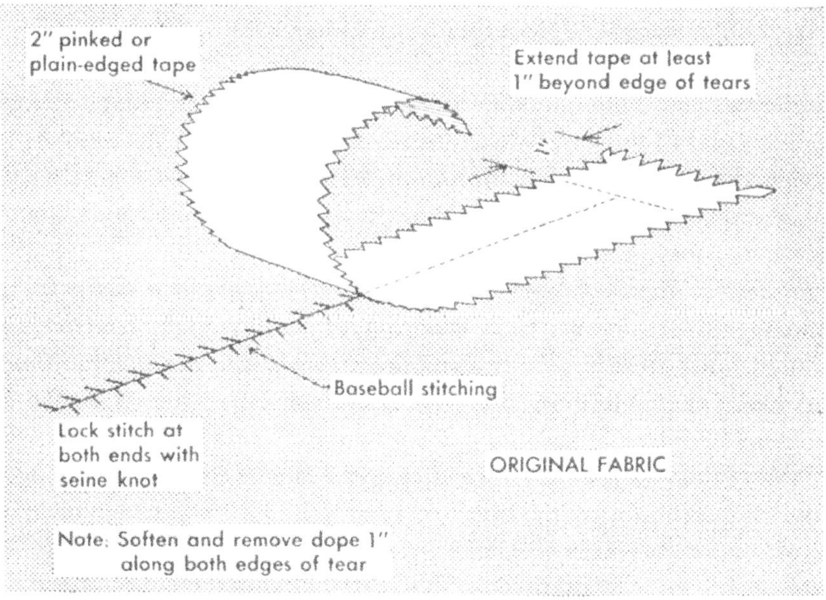

Fig. 172. Patching large tears more than 12 in. long.

of the wing is removed for at least 1 in. in all directions from the edge of the hole. A patch extending at least 1 in. in all directions from the hole is applied in the same manner as in Problem 1.

Problem 4. Repair Damage to Fabric on Wing Tip. The wing tip is very often damaged by coming into contact with the hangar wall, hangar

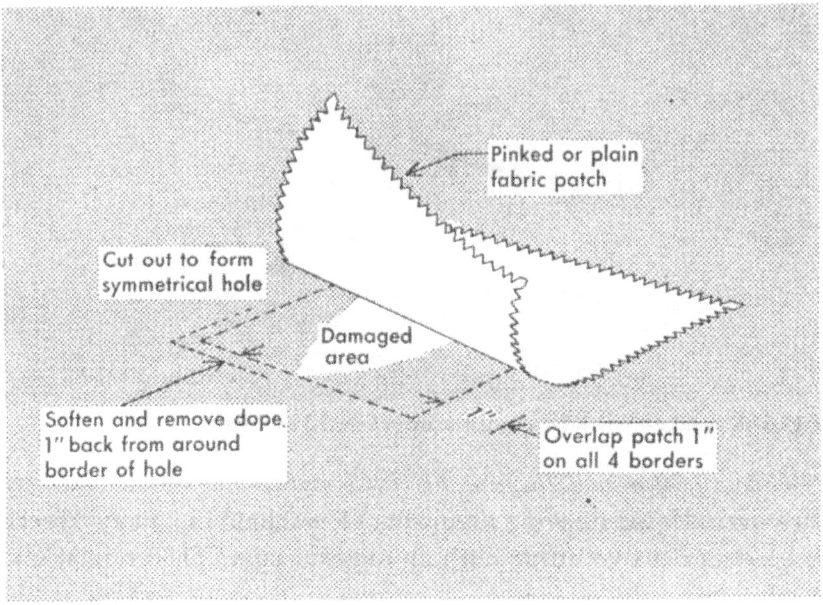

Fig. 173. Patching small V tears less than 6 in. long.

doorframes, etc. This damage may be repaired by removing the tape from the end of the wing for a distance of not less than 1 in. beyond the damaged area. Remove the finish for at least 1 in. around the damaged area by use of dope thinner or other solvent. Stitch the fabric and recover with tape, doping in place as directed in Problem 1.

Problem 5. Repair Damage to Trailing Edge of a Fabric-Covered Wing. Remove the reinforcing tape covering the trailing edge. Remove the original finish from the fabric, and draw the fabric into place by stitching. Dope the reinforcing tape over the repaired area as directed in Problem 1.

Problem 6. Replace a Large Area of Damaged Fabric. Remove the damaged fabric by cutting out the fabric to within 2 in. of the ribs adjacent to the part to be recovered. Remove the original finish from the fabric for a distance of not less than 2 in. beyond the rib adjacent to the repair. In repairs of this type the fabric should extend not less than 3 in. beyond

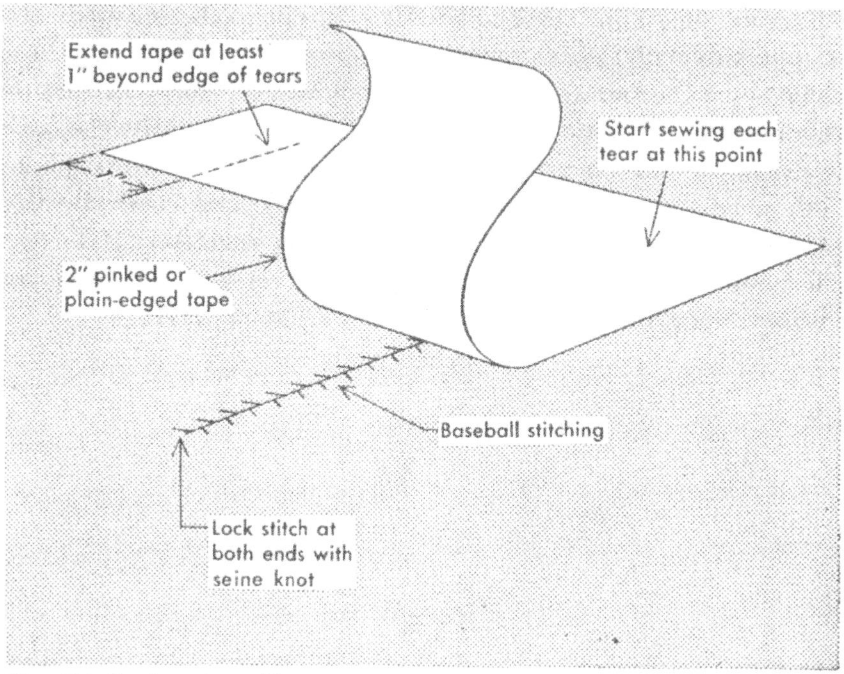

Fig. 174. Patching large V tears more than 6 in. long.

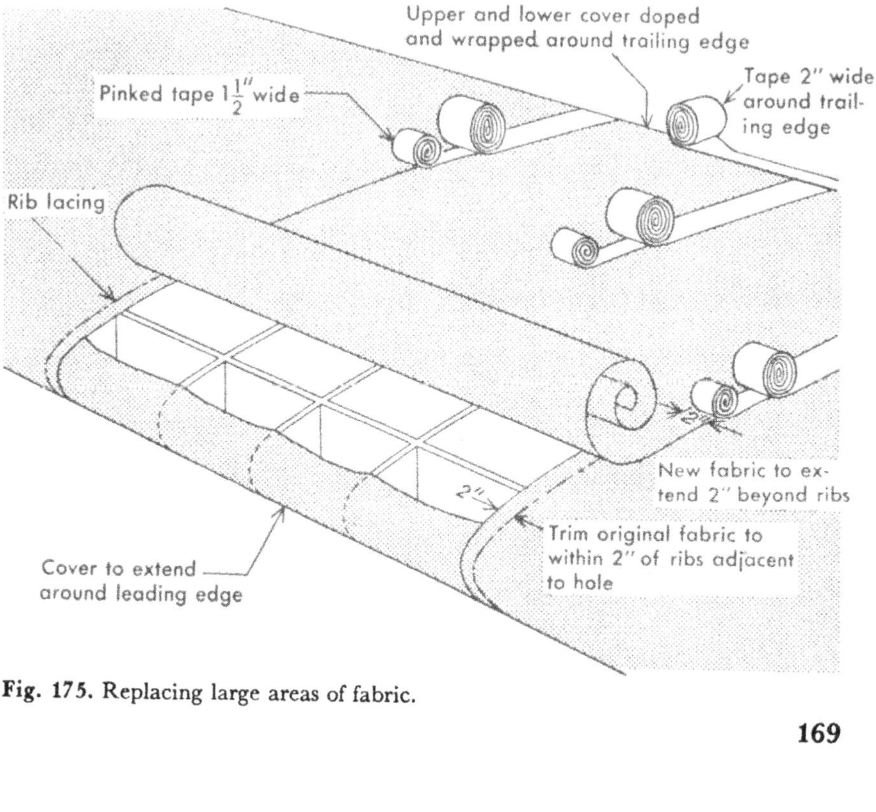

Fig. 175. Replacing large areas of fabric.

the leading edge, 2 in. beyond the ribs adjacent to the repaired area, and approximately ½ in. around the trailing edge. The fabric is then stretched over the area to be repaired, as shown in Figure 175, and the edges doped to the fabric already in place on the wing. After the edges have dried, the new piece of fabric is rib-stitched to the ribs beginning with those adjacent to the repaired area, each rib being covered with a reinforcing tape in the same manner as applying new fabric. The edges of the newly applied fabric should be covered with pinked tape, and the whole doped and finished as directed in Chapter XIII.

XXV APPROVED PLYWOOD REPAIRS

Problem 1. Emergency Repairs to Small Holes in Plywood. An emergency patch may be applied to a small hole in a plywood wing covering in

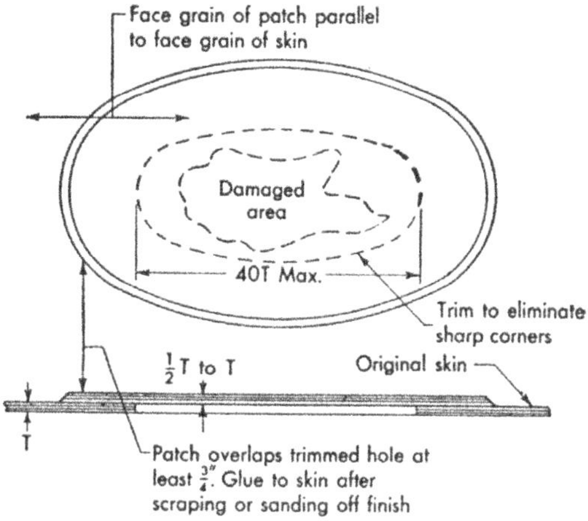

Fig. 176. A temporary patch to a plywood covering.

Fig. 177. Removing the finish from plywood.

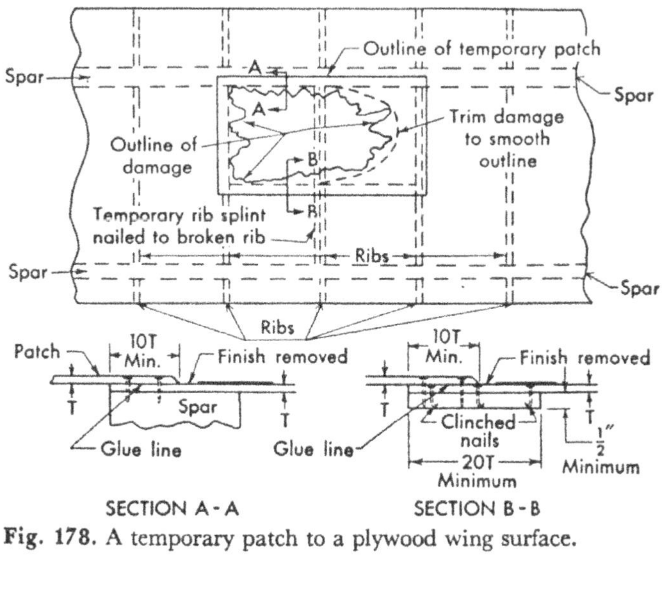

Fig. 178. A temporary patch to a plywood wing surface.

the following manner. With a sharp knife or other suitable tool, trim out the damaged wood to eliminate any sharp corners. Scrape the finish off the plywood to a distance of approximately 1 in. from all sides of the hole. Prepare a plywood patch of such size that it will extend not less than ¾ in. on all sides of the hole. Bevel the edge of the patch. The grain

Fig. 179. A damaged leading edge.

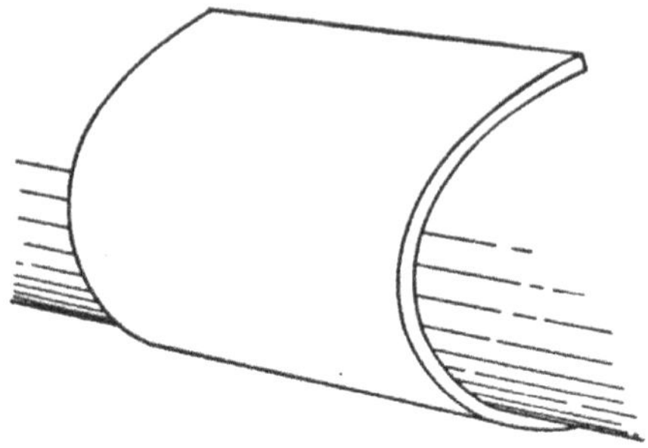

Fig. 180. A temporary patch on a leading edge.

of the patch should be parallel to the grain of the plywood being repaired. The patch may now be glued in place, pressure being applied by a block, if the surface is level, or by means of a sandbag on curved surfaces. A piece of paper should be placed over the patch to prevent the sticking of the glue to the block or sack used to apply pressure.

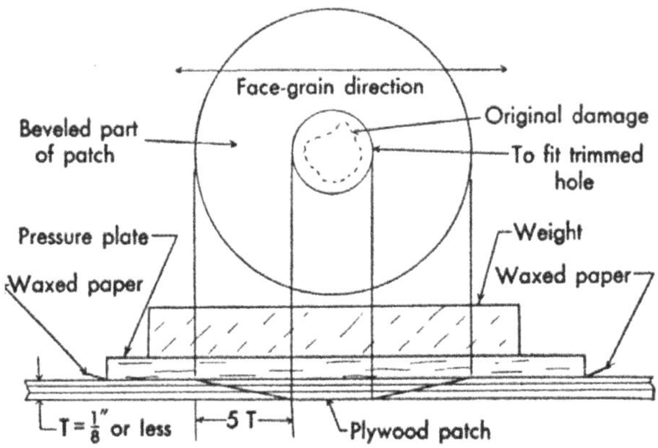

Fig. 181. A splayed patch.

Problem 2. Emergency Repairs to Large Holes. Emergency repairs to large holes in a plywood covering may be made as shown in Figure 178. When the leading edge is damaged, the damaged plywood should be cut out until the edge of the patch will rest on two adjacent ribs. The patch should extend around the leading edge, both top and bottom,

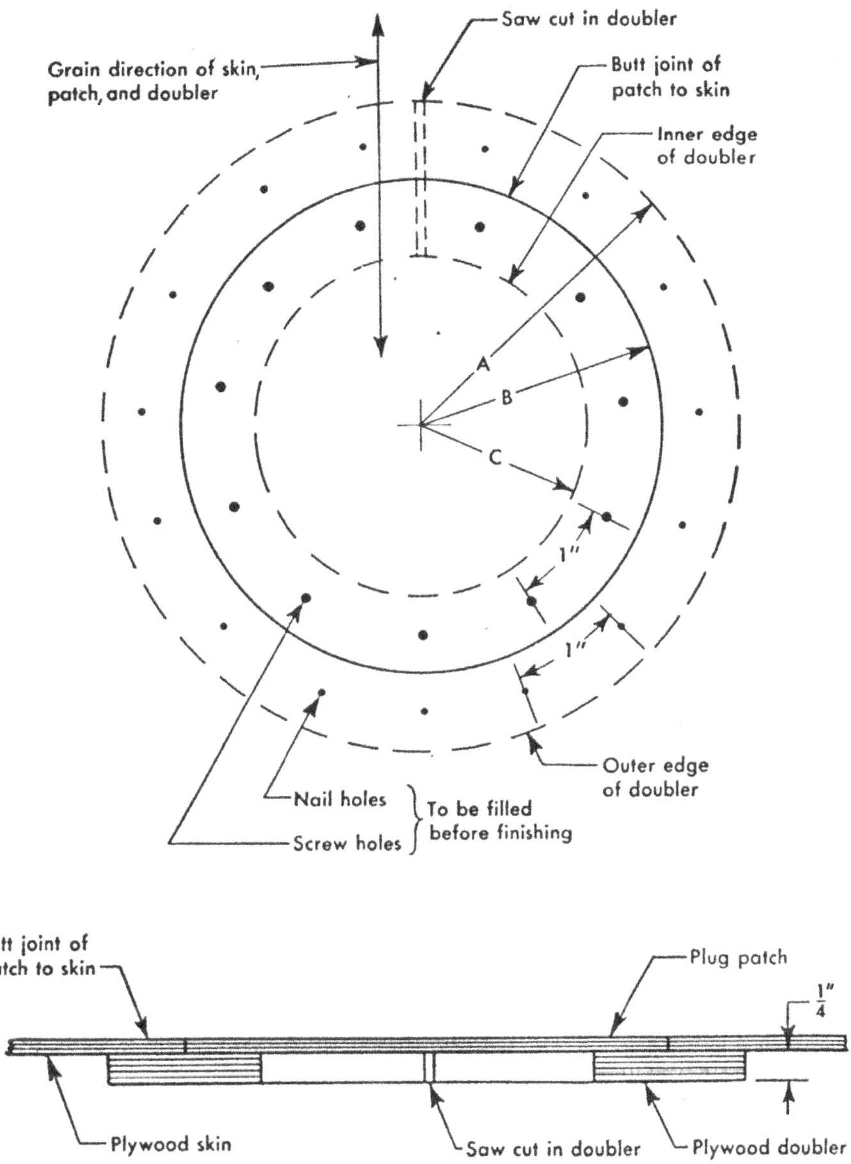

Fig. 182. A plug patch. Dimension (inches)

	A	*B*	*C*
Small circular plug patch	2⅝	2	1⅜
Large circular plug patch	3⅞	3	2⅛

(Two rows of screws and nails required for large patch)

for a distance of not less than 6 or 8 in. The patch should then be nailed and glued into place, soaking the patch if necessary. Use the leading edge for a form to shape the patch, if the plywood has been soaked. Allow to dry before gluing and nailing into place. It is often advisable to dope a piece of fabric over a temporary patch of this kind. Figures 179 and 180 show the steps in making this type of repair.

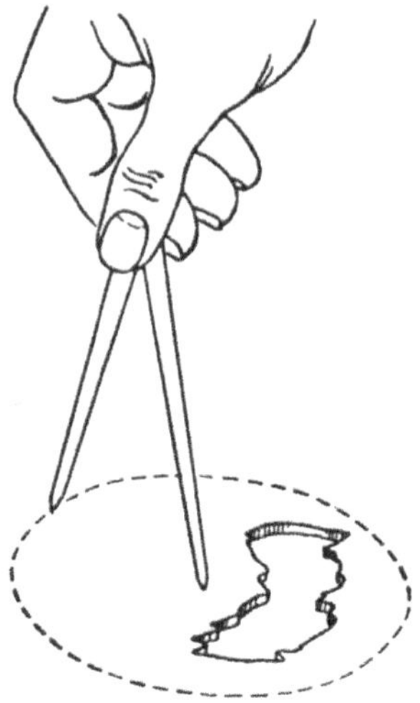

Fig. 183. Prepare to cut out the damaged area in the form of a circle.

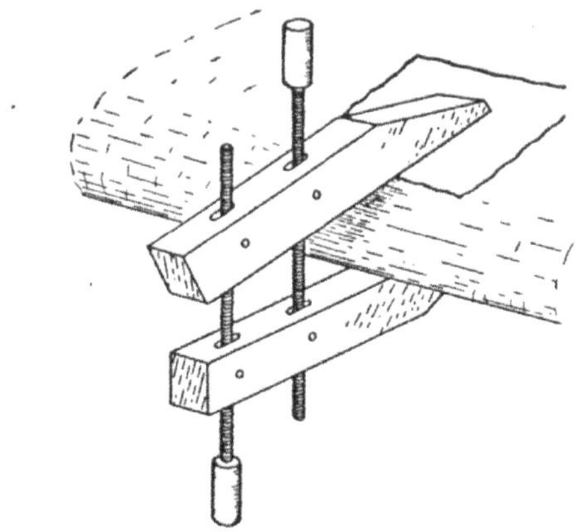

Fig. 184. A patch on the leading edge of a wing covered with paper and clamped.

Problem 3. Permanent Repair to Small Holes. Small holes may be repaired permanently by either a splayed patch or a plug patch. A splayed patch is a patch with tapered edges, but the slope is steeper than a scarfed edge. In a splayed patch the taper is approximately 5 times the thickness of the material. Holes repaired by this method should not exceed in diameter 20 times the thickness of the skin. The hole is cut out in the form of a circle, as shown in Figure 183. The edge of the hole is then tapered evenly for a splayed patch, using either a sharp knife, chisel, or a wood rasp. If a plug patch is used, the edge of the hole is trimmed at right angles to the surface of the plywood, as shown in Figure 182. The edge of the patch is then trimmed to fit the hole exactly and glued in place. The entire patch should be covered with waxed paper or some suitable material to keep the glue from sticking to the clamps or weights used to apply pressure.

An oval plug patch may be used for holes up to approximately 5 to 7 in. in size. The holes should be trimmed to the correct oval shape, and the finish removed from both the inside and outside of the plywood

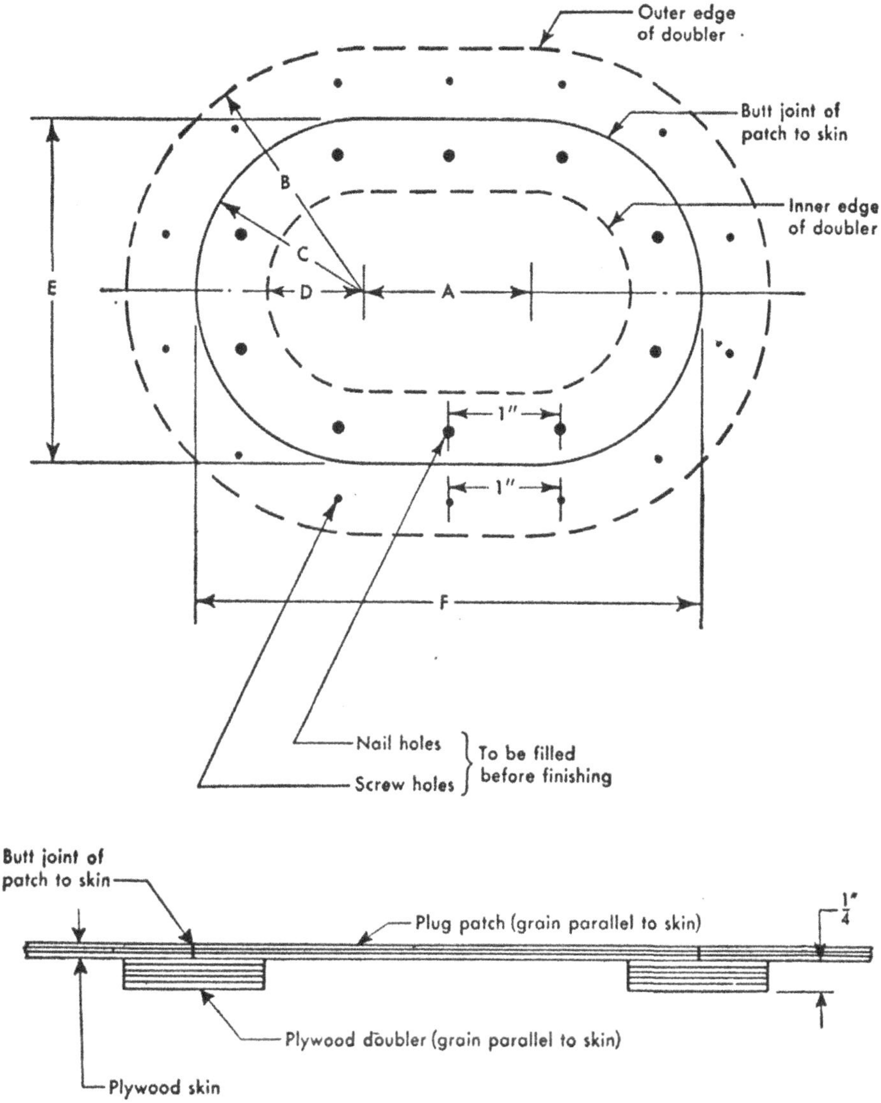

Fig. 185. An oval patch. Dimensions (inches)

	A	*B*	*C*	*D*	*E*	*F*
Small oval plug patch	1½	2⅛	1½	⅞	3	4½
Large oval plug patch	2	3⅜	2½	1⅝	5	7

(Two rows of screws and nails required on large patch)

Fig. 186. Trimming a hole for an oval patch.

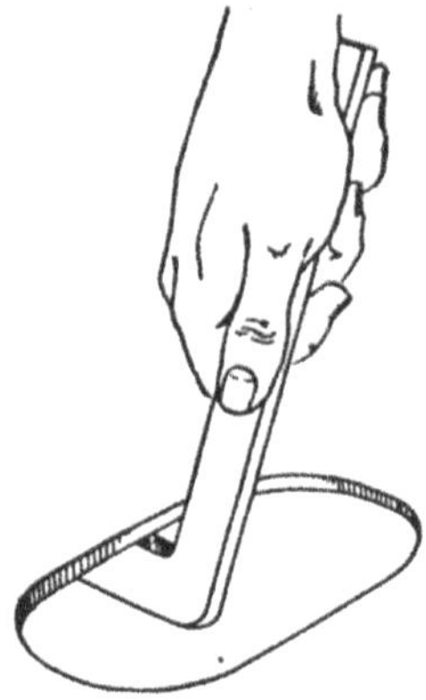

Fig. 187. Removing the finish from the under side of the plywood with a scraper.

Fig. 188. Applying a doubler for an oval patch.

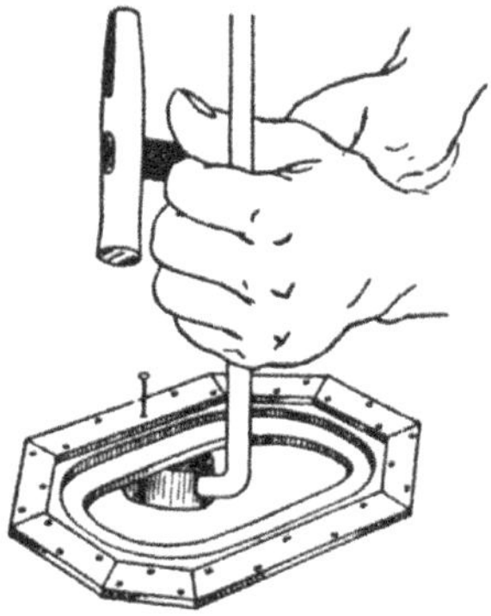

Fig. 189. Nailing strips used to hold a doubler in place.

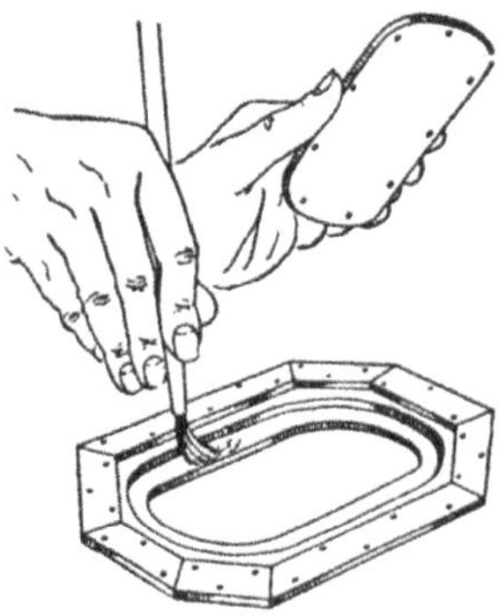

Fig. 190. Applying glue for an oval patch.

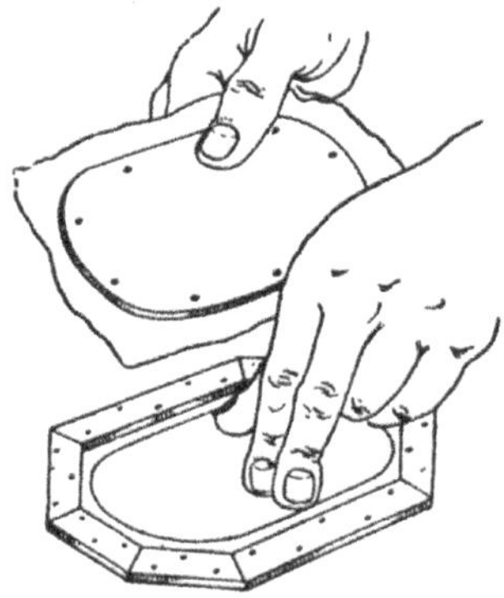

Fig. 191. An oval patch in place. The pressure plate is in the background.

covering for at least 1 in. from the edge of the hole, as shown in Figures 185 and 186. In applying the oval patch, a doubler is used. A doubler is a reinforcement placed under the edge of the hole inside the skin. This doubler is nailed and glued in place using nailing strips. The patch

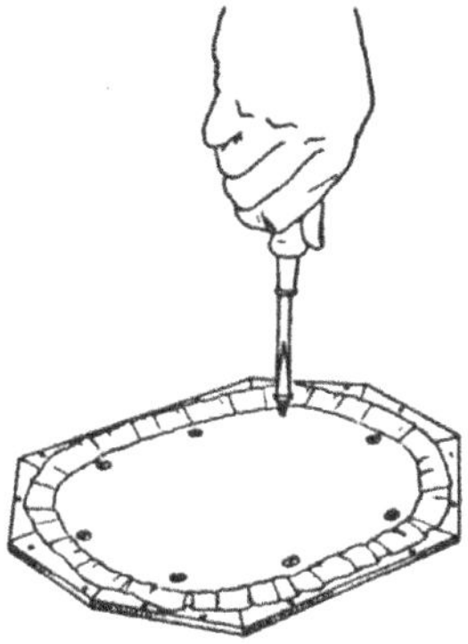

Fig. 192. Screwing a pressure plate into place.

Fig. 193. Removing excess glue with a scraper after the pressure plate and nailing strips have been removed.

Fig. 194. Screw holes filled with plastic wood and sandpapered smooth.

is cut to the exact size of the hole and glued and screwed to the doubler, using a pressure plate. On larger sized patches, two rows of nails or screws should be used. After the patch has thoroughly dried, remove the nailing strips with a pair of pincers. Scrape off the excess glue and finish smooth with sandpaper. Figure 196 shows the placing of a doubler for a round plug patch by first splitting the doubler on one side.

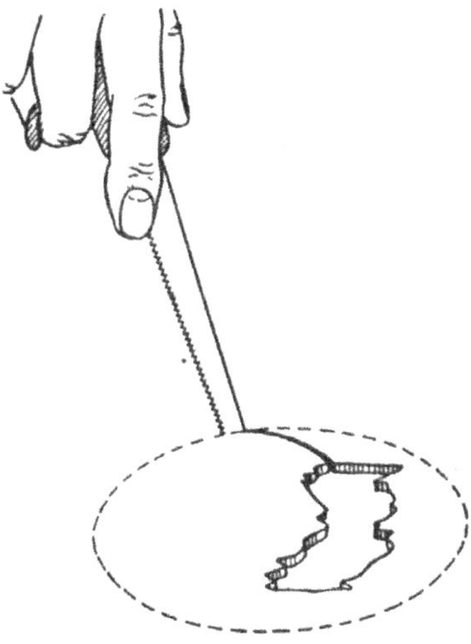

Fig. 195. Sawing out a damaged area in the form of a circle.

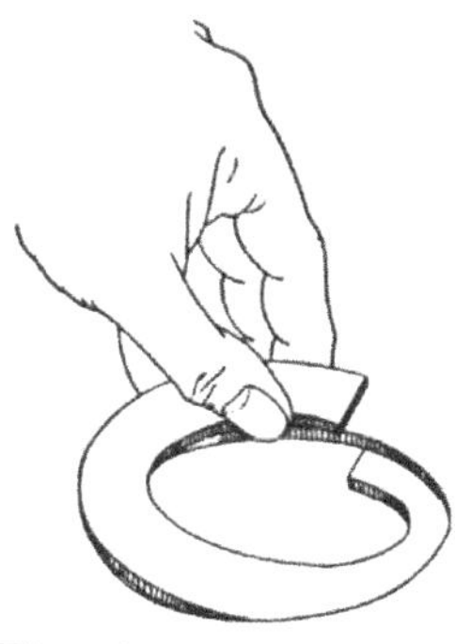

Fig. 196. A split doubler being put in place under the plywood skin.

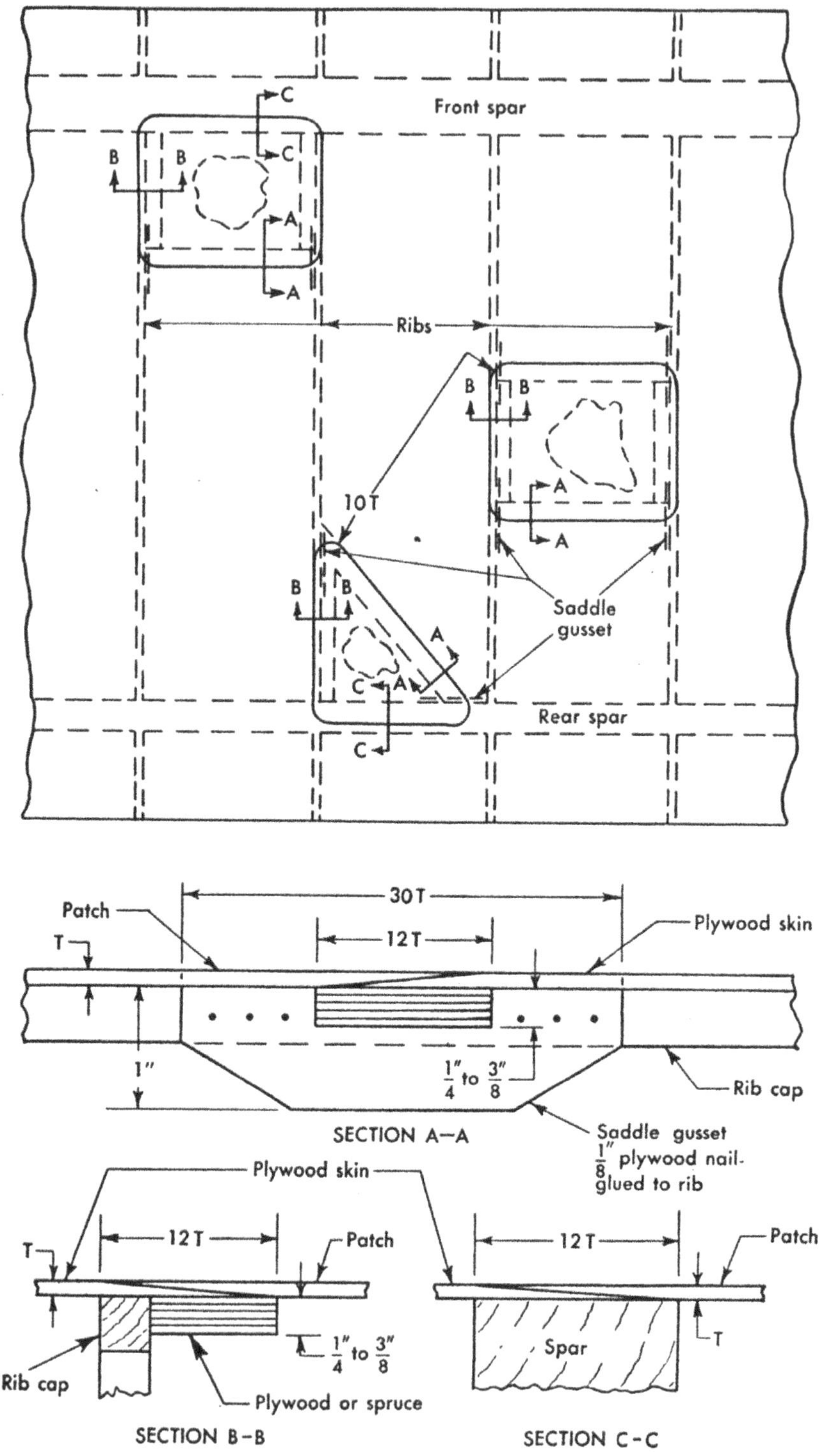

Fig. 197. Scarf patches on a wing surface.

Problem 4. Scarf Patches. Damaged areas too large to be covered by oval or round patches should be repaired by inserting a scarf patch. Scarf patches may be used in any place on the plywood skin except that no horizontal scarf joint is allowed on the leading edge. In sharply

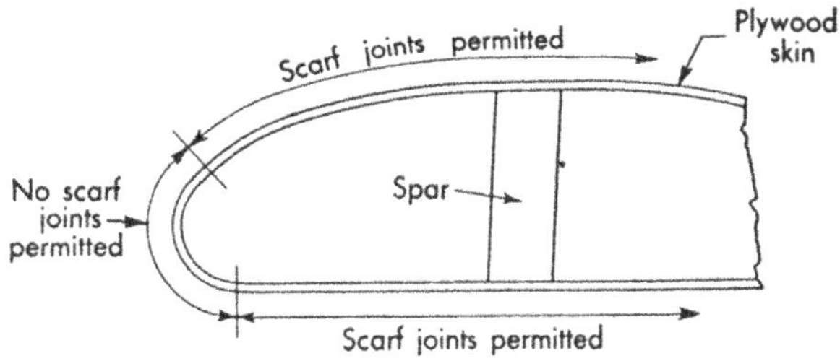

Fig. 198. No scarf joints are allowed on the leading edge of a wing.

curved parts it is necessary to preform the plywood to be used before scarfing its edges. Patches covering an area extending over two or more ribs should have reinforcing strips placed along the edge of the ribs, as shown in Figure 199. The edge of the plywood may be scarfed by a hand plane, scraper, accurate sanding block, or chisels. The slope used

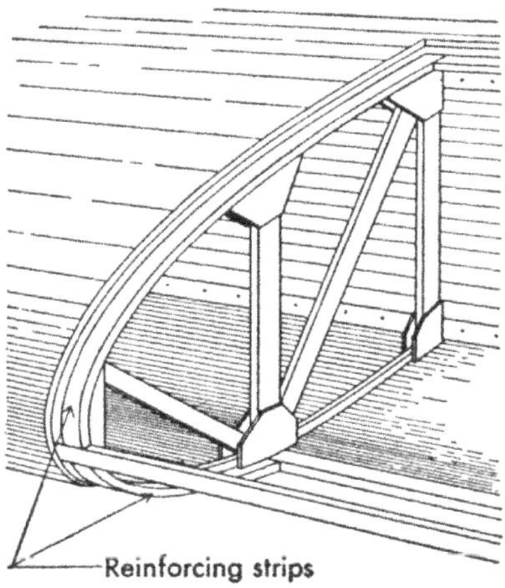

Fig. 199. Reinforcing strips along the edge of the ribs.

on a scarf patch should not be greater than 1 in 15. The corner of a hole for a scarf patch should be rounded, and proper backing strips should be placed under the edges of the plywood skin. Nailing strips should be used to fasten the backing strips into place. Backing strips should be made of a softwood such as spruce or poplar. The edges of the scarfed end of the patch are glued and nailed into place using hot

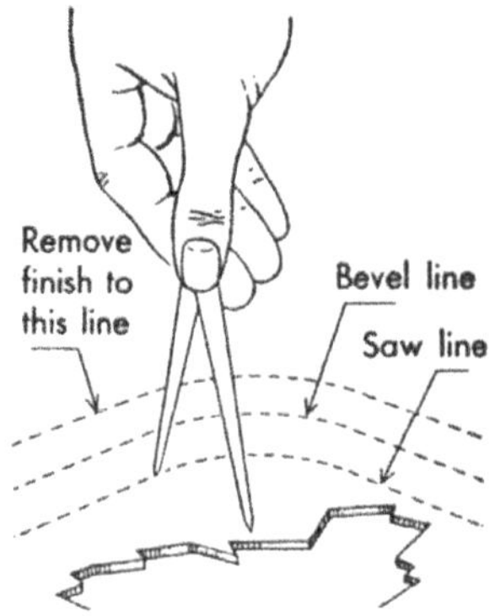

Fig. 200. Rounding the corners for a scarf patch.

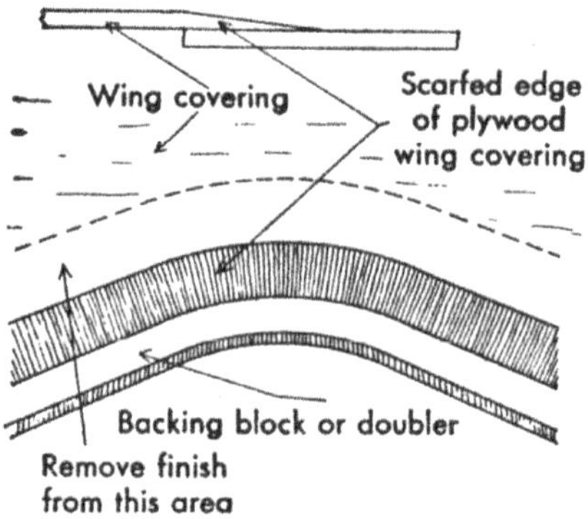

Fig. 201. Backing strips for a scarf patch.

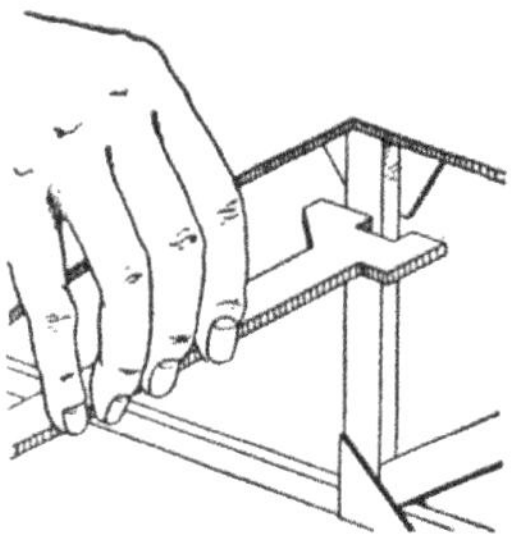
Fig. 202. Backing strips to hold the scarf patch in place.

Fig. 203. Hot sandbags used to apply pressure.

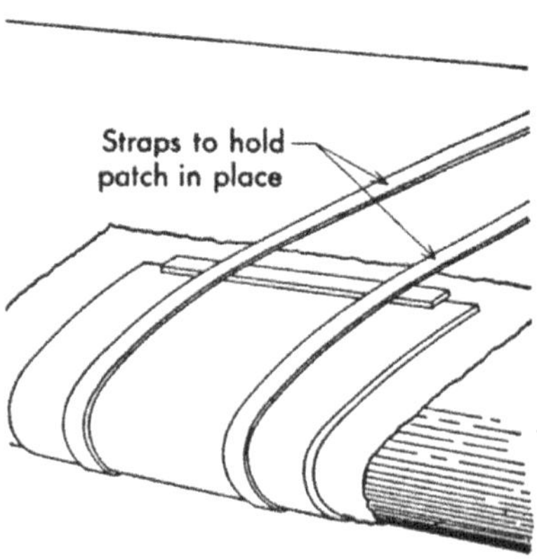

Fig. 204. Leading edge used to form a patch.

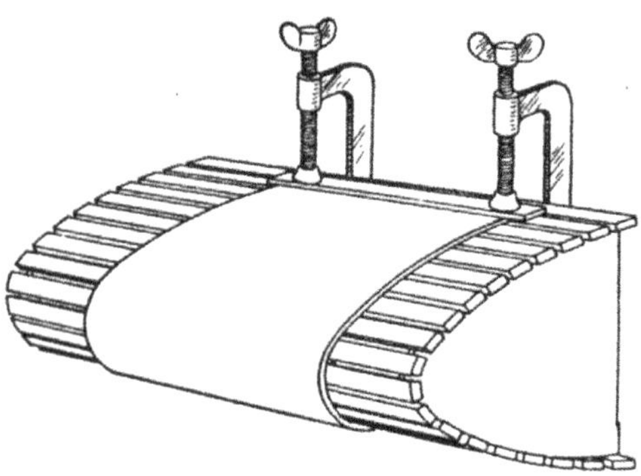
Fig. 205. Forming a leading-edge patch.

sandbags to apply pressure. Plywood for a patch on the leading edge should be preformed either by using the leading edge of the wing as a form, or as shown in Figure 205. Place heavy burlap or paper under the plywood to prevent injury to the leading edge. Backing strips for the

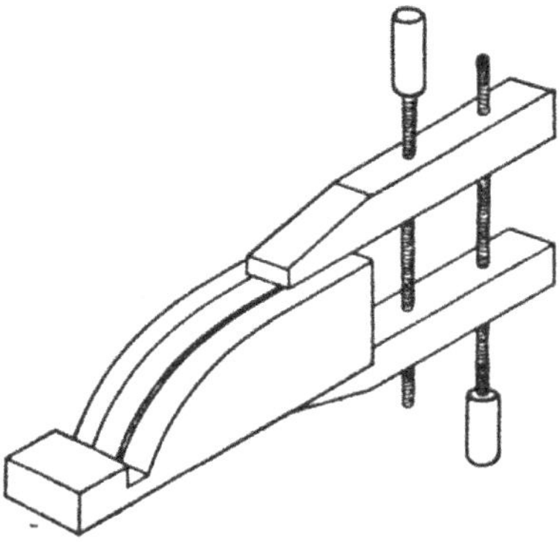

Fig. 206. A hand clamp used to form a backing strip.

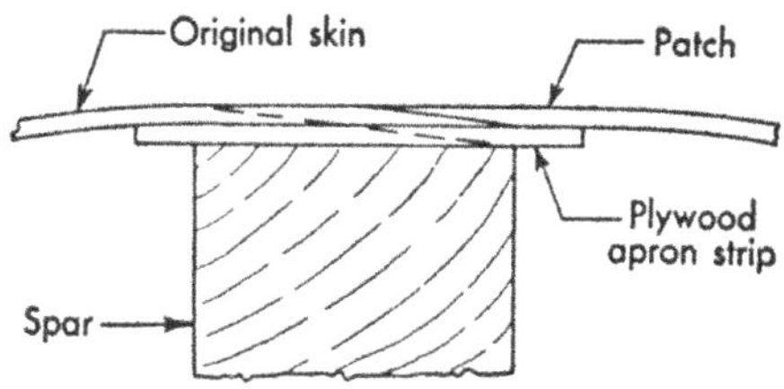

Fig. 207. A scarf joint over an apron strip.

leading edge may be formed by using hand clamps as shown in Figure 206. After the patch has thoroughly dried, nailing strips should be removed with a pair of pincers and the surface properly finished by scraping off excess glue and finishing with sandpaper. Figure 207 shows the repair of a scarf joint over the apron strip.

XXVI APPROVED WING–STRUCTURE REPAIRS

The wing framework consists of spars, ribs, leading-edge strip, wing-tip bow, trailing-edge strip, compression members, braces, and reinforcing members. A scarf joint is the most satisfactory method of making a repair to a solid wood member. The joint should be cut accurately and the same slope be made on each of the pieces to be joined. For the proper method of making a scarf joint, refer to Chapter XI.

Problem 1. Repair a Solid or Laminated Rectangular Spar. A spar may be spliced at any point except under wing fittings or where other parts, such as landing-gear fittings and engine fittings, are attached to the spar. Reinforcing plates must be used on all scarf repairs to spars. The proper slope of a scarf joint for a solid or rectangular spar should not be less than 10 times the thickness of the part spliced. Figure 208 shows how

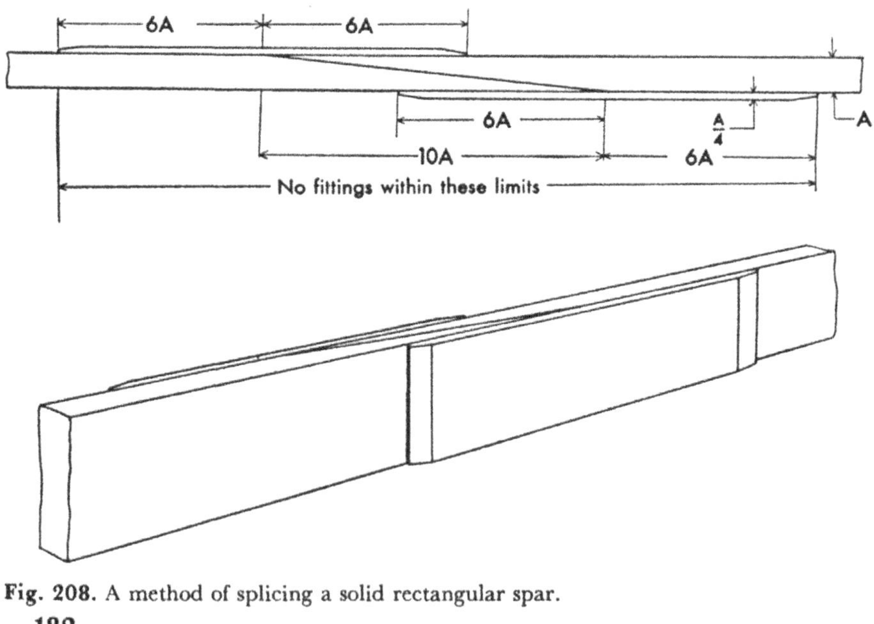

Fig. 208. A method of splicing a solid rectangular spar.

this splice should be made. This problem consists of the removal and splicing of the front spar in the root-bay section constructed in Chapter XV. Either end of the spar may be replaced. The fabric covering of the wing should be loosened along the trailing edge and completely removed from the wing. The wing walk should be removed. The plywood leading edge should be loosened from the front spar. All fittings should be loosened and removed from the front spar. The blocks on top of the front spar are next removed by the use of a chisel and hammer. The ribs are loosened from the front spar by removing the gussets at the top and bottom of the vertical-rib members. Using either a knife blade or a putty knife, the root of the front spar is loosened from the root rib and may be removed by sliding outward. After the splice has been made in the spar, as shown in Figure 208, the spar should be replaced and the wing rebuilt as before. The vertical members of the ribs are replaced, making close contact with the reinforcing blocks, and new gussets glued and nailed into place.

Problem 2. Repair Cap Strips and Ribs. A cap strip of a wood rib may be repaired by means of a scarf joint which is reinforced on the side opposite the wing covering by a spruce block which extends beyond the scarf joint not less than 3 times the thickness of the strips being repaired. The entire joint, including the reinforcing block, is reinforced on each side by a plywood reinforcing as shown in Figure 209. When the cap strip is to be repaired at a point where there is a joint between it and cross members of the rib, the repair is made by reinforcing the

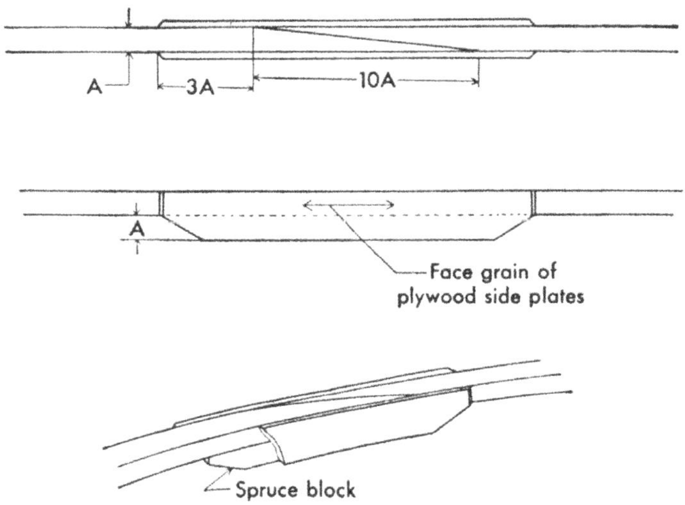

Fig. 209. A rib cap-strip repair.

scarf joint with plywood gussets, as shown in Figure 210. When it is necessary to make a repair to a cap strip at a spar, the joint should be reinforced by means of a continuous gusset extending over the spar as shown in Figure 211. Edge damage, cracks, or other local damage in a spar may be repaired by removing the damaged portion and gluing in a properly fitted block, as shown in Figure 212, reinforcing the joint by means of plywood or spruce blocks which are glued into place. The

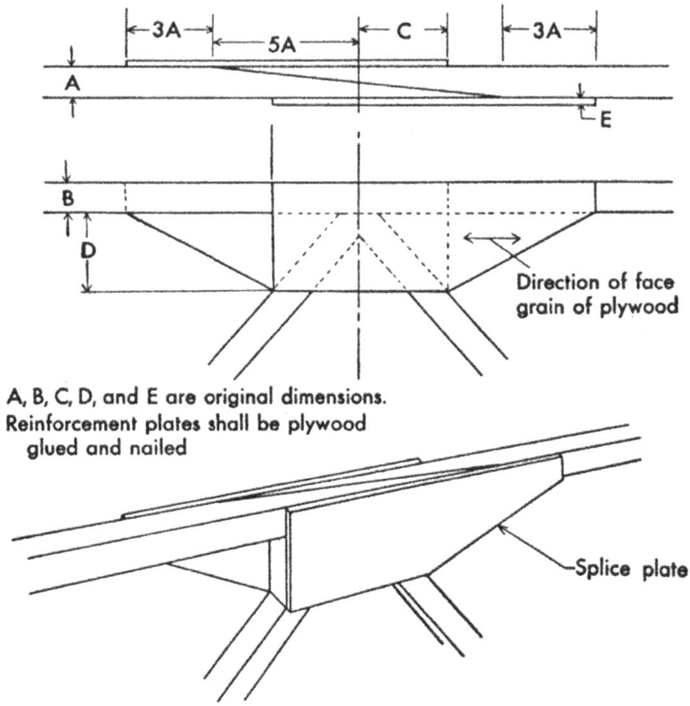

Fig. 210. A rib repair at a joint.

trailing edge of a rib may be replaced and repaired by removing the damaged portion of the cap strip and inserting a softwood block of white pine, spruce, or basswood. The entire repair is then reinforced with plywood gussets and nailed and glued as shown in Figure 213. A compression-rib repair may be made as shown in Figure 214. Compression ribs are of many different types, and the workman must judge the proper method for making a repair to any part of this type of rib. Figure 214 shows a repair made to a compression rib built up of a plywood web and three longitudinal members, the center one of which has been repaired by a properly reinforced scarf joint and an outside layer of plywood.

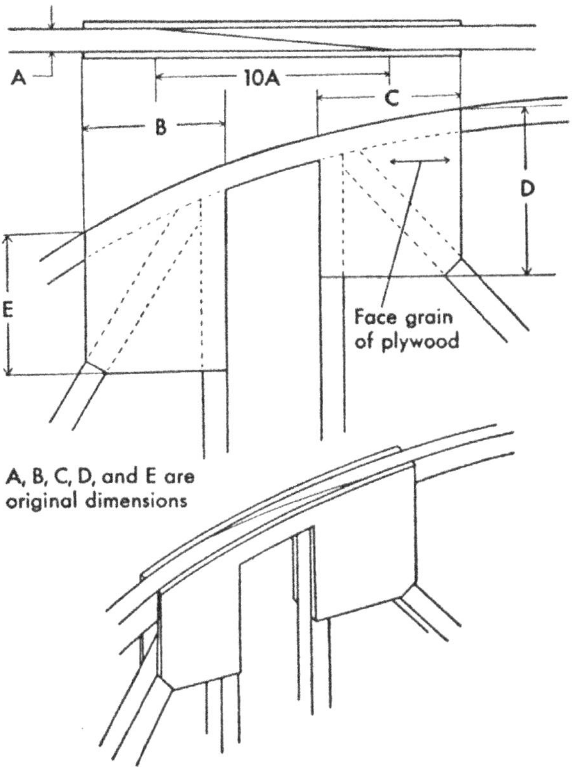

Fig. 211. A rib repair at a spar.

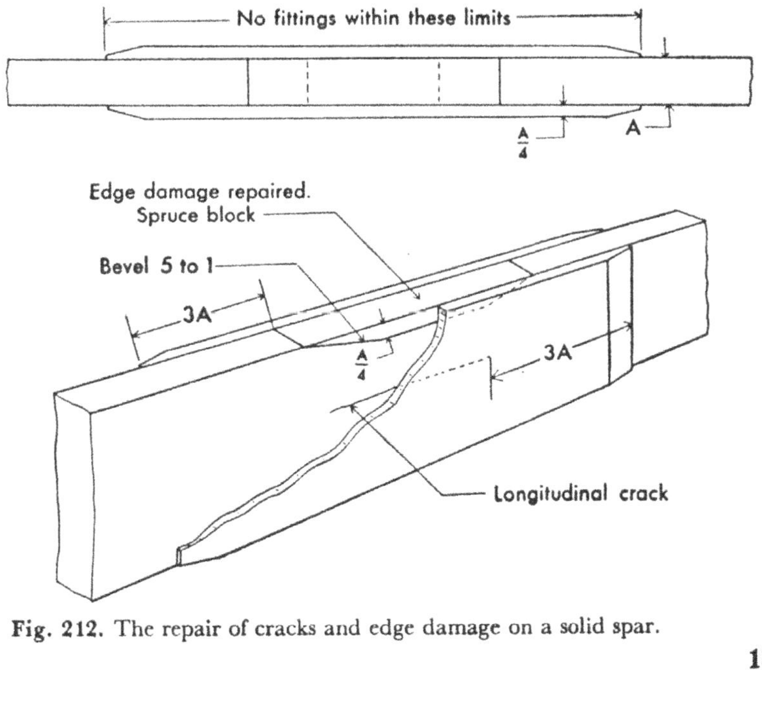

Fig. 212. The repair of cracks and edge damage on a solid spar.

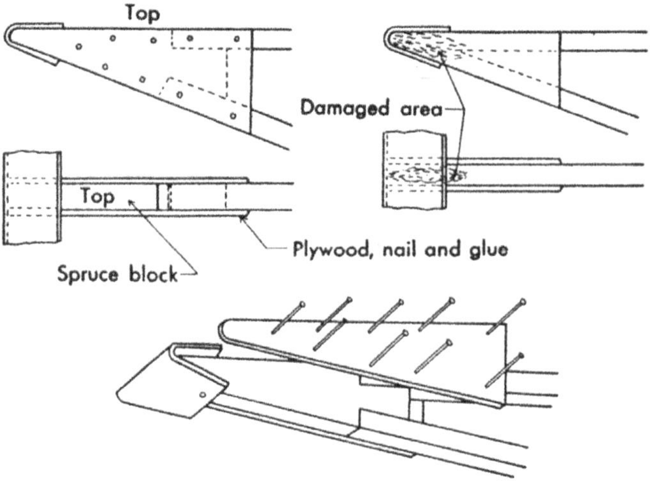

Fig. 213. A rib trailing-edge repair.

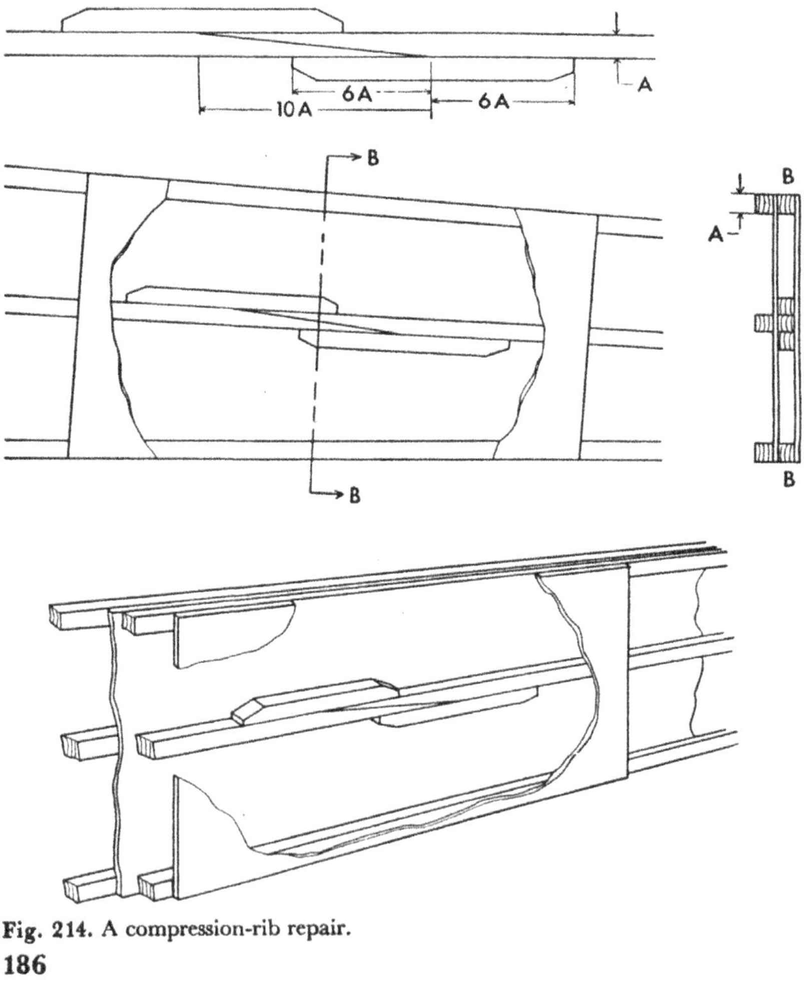

Fig. 214. A compression-rib repair.

Such members as glue blocks, filler blocks, compression members, braces, and rib diagonals should not be repaired, but should be replaced. Wherever it is possible to replace a damaged member, it is always better to do so than to attempt a repair.

Wing-tip bows and other laminated structures, when it is impossible to replace the entire members, may be repaired by sawing out the damaged portion and splicing in by means of a scarf joint a preformed laminated section. A temporary repair may be made by splicing in a solid piece with a scarf joint, if the damage is not too extensive.

Fig. 215. A scarf joint in a wing-tip bow.

Problem 3. Finish Repairs Properly. After repairs have been made the proper finish should be applied. When making repairs, care should be taken that the surrounding areas or the area repaired is not covered with glue or any other substance which will interfere with proper finishing. Oil or grease may be removed with naphtha. Marks made by lumber crayons or grease pencils should be removed. Any chips, shavings, sawdust, glue particles, or other waste material should be removed from enclosed spaces. A vacuum cleaner may be used to remove such material. End-grain surfaces and plywood edges should be given two coats of sealer. Drilled holes should be finished with two coats of sealer and one coat of aluminized varnish. Nail heads, screw heads, etc., should be covered with a strip of tape or other material to prevent cracking of the finish. Holes that are left by removing nails when nail-strip gluing has been used should be filled with plastic wood. All new woodwork should be given not less than two coats of sealer. Exposed surfaces should be refinished to match the original finish.

Control-surface repairs are made in the same manner as wing repairs.

XXVII APPROVED FUSELAGE REPAIRS

Longerons, Stringers, Cross Braces, and Vertical Braces are repaired by scarfing and reinforcing. Care should be taken that the grains are properly matched in making these repairs.

Problem 1. Repair Bulkheads in Fuselage. In repairing a bulkhead, a rectangular piece of plywood skin large enough to allow the repair to

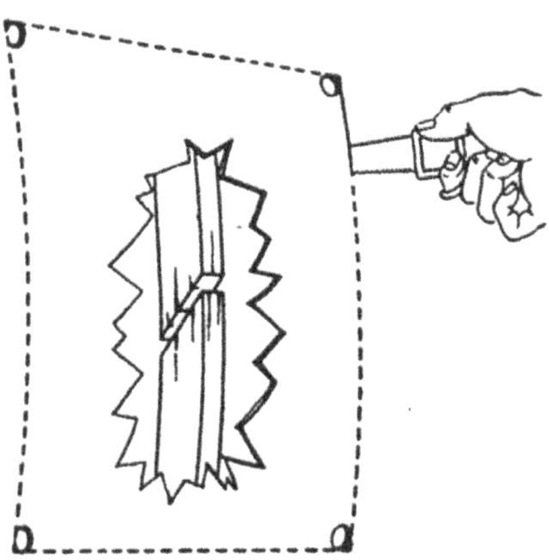

Fig. 216. Removing damaged material from a plywood fuselage.

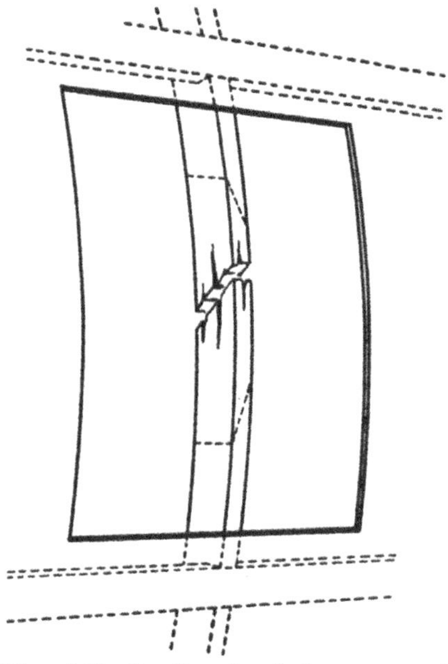

Fig. 217. Scarf ends of the structural member.

be made should be removed, as shown in Figure 216. The bulkhead is then scarfed, and a repair piece prepared of the proper shape. This piece is glued into place, and reinforcing members are applied. A doubler is placed around the edge of the hole cut in the plywood, and a scarf patch applied, following the procedure given in Chapter XXV, Problem 4. Figures 219 and 220 show the use of an adjustable curve for reproducing the curved portions of bulkheads or skin stiffeners.

Problem 2. Repair a Skin Stiffener. Skin stiffeners are repaired either by scarf splicing in a part to replace the damaged material, or by removing that portion of the skin stiffener between bulkheads and replacing as

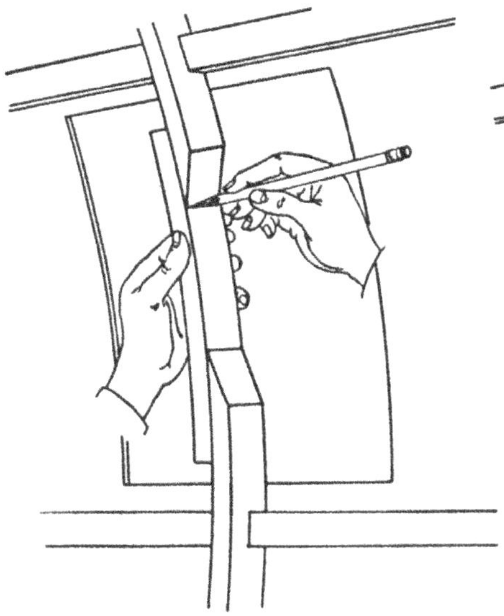

Fig. 218. Marking a piece of material to be used in repairing a bulkhead.

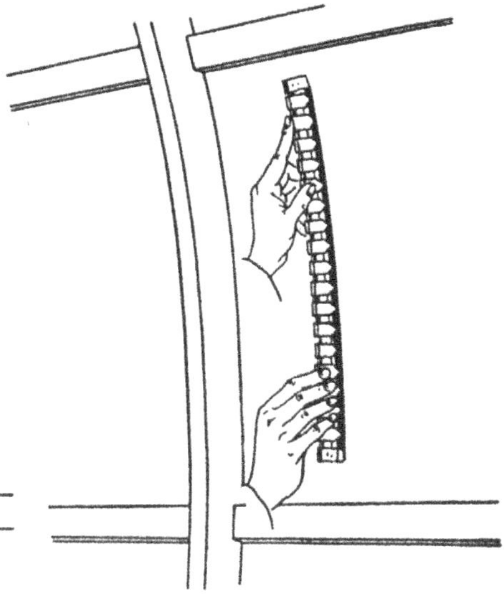

Fig. 219. Using an adjustable curve to get the curve of the fuselage covering.

shown in Figure 221. The ends are strengthened by the use of saddle gussets.

Problem 3. Repair and Replace Bearing Blocks and Plywood Reinforcements. Bearing blocks and plywood reinforcements are repaired or replaced

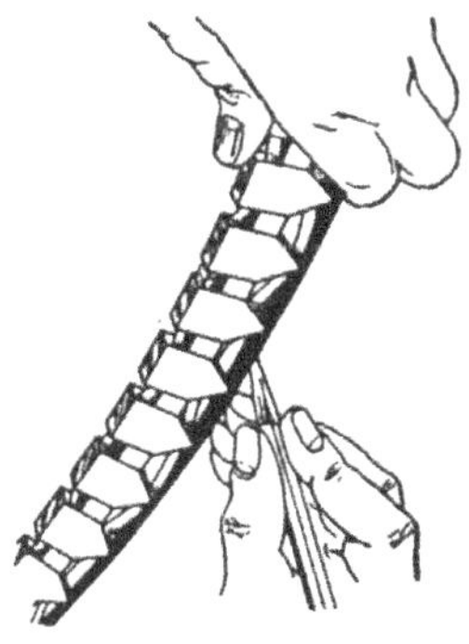

Fig. 220. Using an adjustable curve to mark material for the repair of a bulkhead.

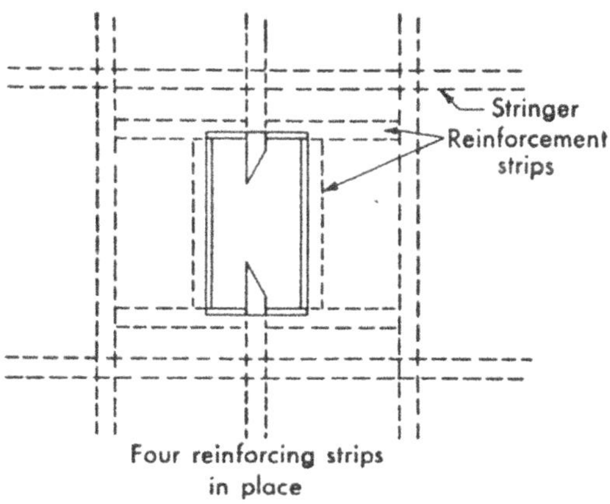

Fig. 221. Repair to a skin stiffener.

in the same manner as described in Chapter XXVI for wing-structure repairs.

Problem 4. Repair Fairings. Fairings do not carry structural loads. Damaged material may be replaced by gluing in filler blocks of the same material, or the entire fairing may be replaced.

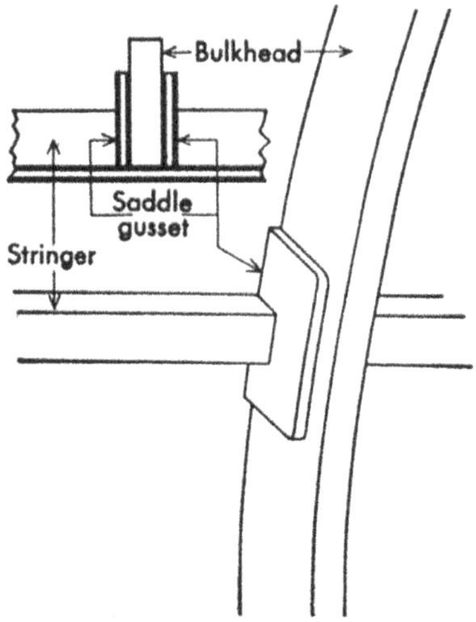

Fig. 222. A saddle gusset to reinforce a skin stiffener.

XXVIII MISCELLANEOUS PROBLEMS

The construction used in these problems is typical of construction used in aircraft woodwork.

Problem 1. Construct a Section of a Vertically-Laminated Spar. The spar is built up of five laminations, three vertical grain and two flat-sawed pieces. The grain should be matched as shown in Figure 55. A section 30 in. long may be constructed. The laminations will be $\frac{1}{4}$ in. thick and 6 in. wide. The fit between the pieces should be as nearly perfect as possible. Hardwood backing blocks should be prepared, each 1$\frac{1}{4}$ in. in thickness and of sufficient size to cover the material being glued. A pressure of approximately 150 lb. per sq. in. should be applied.

Problem 2. Splice a Vertically-Laminated Spar. The section of the spar built in Problem 1 may be used for this problem. A scarf joint is prepared and glued as shown in Chapter XI. After gluing, the joints are to be reinforced with plywood which is glued and nailed as shown in Chapter XXVI, Problem 1. All splices in spars should be reinforced. The plywood reinforcing on each side should be $\frac{1}{4}$ the thickness of the spar.

Problem 3. Build a Section of an I-Beam Laminated Spar. This portion of a spar should be 8 in. deep, and the web formed from two laminations, edge grain, each $\frac{5}{16}$ in. thick. Each of the top and bottom flanges will be vertically laminated of four pieces $\frac{3}{8}$ in. thick and 1 in. deep. (See Figure 57.) This section of the spar is approximately 30 in. in length. The two pieces forming the web are prepared and glued together. After the web is thoroughly dry, the spar flanges are prepared and glued to the top and bottom of the web in a single operation.

Problem 4. Splice an I-Beam Spar. The I-beam spar constructed in Problem 3 is used for this problem. A scarf joint is prepared and glued, and the proper reinforcing blocks are glued and nailed into place. (Chapter XXVI, Problem 1.)

Problem 5. Build a Section of a Plywood Box Spar. This spar is to be 8 in. deep, 4 in. wide, and approximately 30 in. long, built from $\frac{3}{32}$-in.

plywood with corner blocks 1 ⅛ in. deep and 1 in. wide, edge grain on the wide surface. The plywood for the two sides is cut to the proper shape, which will be 8 in. wide and 30 in. long. The surface of the plywood that is to be glued to the corner blocks should be lightly sanded to remove any gloss or glue on the surface of the plywood. The corner blocks are now prepared to exact size, and the plywood forming the sides of the spar is nail-glued to these blocks. There should be two rows of nails approximately ¾ in. apart and the nails in the rows staggered. The nails should penetrate the block not less than 3 times the thickness of the plywood. The nails should be driven down firmly but with the heads not embedded in the wood. After allowing each side to dry not less than 8 hr., the top and bottom webs are nail-glued into place.

Problem 6. Construct a Cap-Strip Bending Form. In making repairs to the nose portion of damaged ribs, it is not always desirable to build a complete jig when replacing a small portion of the cap strips of the rib. It is difficult to replace this portion of the rib without a jig. The top and bottom cap strips should be bent before making the repair. Figure 206 shows the method of construction of this bending form, using a shaped block and hand clamp

Problem 7. Prepare Three Cap Strips for Repairs. Using the cap-strip bending form prepared in Problem 6, prepare three cap strips for repairs to the nose section of the ribs used in the root bay already constructed. Prepare cap strips of the same material and the same size as the cap strips used in the ribs. Soak one end of these strips in hot water until flexible, 30 min. should be sufficient, and place side by side in the cap-strip bending form. Tighten on the form by means of two C-clamps and a cross piece. Just enough pressure should be exerted to get a snug fit on the form. Excessive pressure may deform the strips. Allow to dry approximately 24 hr. Use these cap strips in making repairs to the ribs.

Problem 8. Install an Inspection Plate in the Plywood Covering of the Wing. Install an inspection plate in the plywood covering of the wing tip which has already been constructed. The method for installing this plate is the same as for installing a plug patch except that the cover is not glued into place. The plate is fastened into place with screws provided with plate nuts.

Problem 9. Install an Inspection Plate in a Fabric-Covered Wing. Install an inspection plate at the inner corner of the outer compression rib of the front spar in the root bay constructed in Chapter XV. Prepare a frame of spruce or other suitable material. Cut the proper-sized hole in the fabric, turning back the fabric over the spar and rib. Glue and nail the inspection frame into place. Prepare an inspection plate of $\frac{3}{64}$-in. plywood and fasten into place on this frame with screws having plate nuts which have been installed on the frame. Bring the fabric to the edge of the inspection opening and stitch to the inspection hole frame. Dope into place and reinforce with tape.

Problem 10. Using Wing-Tip-Bow Jig, Bend a Solid Piece of Oak or Birch and Form a Wing Bow.

TYPICAL REPAIR PROBLEMS

Fabric repairs should be made as directed in Chapter XXIV.

Problem 1. Repair a 3-in. tear in the fabric covering of a wing.
Problem 2. Repair a 3-in. V-tear in fabric covering.
Problem 3. Repair a hole 1 × 2 in. in fabric covering.
Problem 4. Repair a hole 6 × 10 in. in fabric covering.
Problem 5. Repair a hole 3 × 5 in. through both top and bottom fabric at the trailing edge of a wing.
Problem 6. Replace damaged fabric extending over three adjacent ribs.
Problem 7. Recover wing tip from last rib to tip of wing.
Problem 8. Remove and replace rib stitching along one rib between spars replacing reinforcing and pinked tape.
Problem 9. Repair hole in fabric 5 × 7 in. in leading edge.
Problem 10. Repair fabric which has been ripped along edge of wing walk.

Plywood repairs are made as directed in Chapter XXV.

Problem 1. Repair hole 1 × 2 in. in plywood wing covering with temporary patch of doped-on fabric.
Problem 2. Repair hole 3 × 5 in. in plywood wing covering with temporary plywood patch.
Problem 3. Repair hole 1 × 1½ in. in plywood with a round splayed patch.
Problem 4. Repair hole 1½ × 2 in. in plywood with round plug patch.
Problem 5. Repair hole 2½ × 3 in. in plywood with oval plug patch.
Problem 6. Repair hole 3½ × 4 in. in plywood with oval splayed patch.
Problem 7. Repair hole 12 × 14 in. damaging one rib in leading edge with temporary patch.
Problem 8. Repair hole 5 × 7 in. in plywood with oval scarfed patch.
Problem 9. Repair hole 5 × 7 in. in plywood with square scarfed patch.
Problem 10. Repair damaged plywood leading edge extending over two ribs by means of a scarfed patch.

Rib repairs are made as directed in Chapter XXVI.

Problem 1. Replace verticals of one rib on each side of front spar.
Problem 2. Replace two diagonals of one rib between spars.
Problem 3. Replace 3 in. of trailing edge of a main rib.

Problem 4. Replace 3 in. of trailing edge of a compression rib.
Problem 5. Replace top cap strip of main rib.
Problem 6. Replace bottom cap strip of a false rib.
Problem 7. Replace nose block of compression rib.
Problem 8. Make repair to split plywood web in compression rib between spars.
Problem 9. Make scarf splice in cap strip between spars.
Problem 10. Make scarf splice in cap strip over a spar.

Wing structure repairs are made as directed in Chapter XXVI.

Problem 1. Splice rear spar.
Problem 2. Replace plywood plates on rear spar.
Problem 3. Repair hole 3 × 4 in. in plywood walk strip.
Problem 4. Remove and replace abrasive cloth on walk strip.
Problem 5. Remove and replace root compression rib.
Problem 6. Remove and replace front spar.
Problem 7. Remove and replace wing-tip bow.
Problem 8. Splice in a 6 in. section of wing-tip bow while in place.
Problem 9. Remove and replace leading-edge strip in root bay.
Problem 10. Splice in a 6-in. piece of leading-edge strip in wing-tip bay.

Fuselage repairs are made as directed in Chapter XXVII.

Problem 1. Remove and replace a longeron.
Problem 2. Splice in a 6-in. section of a longeron.
Problem 3. Remove one bulkhead.
Problem 4. Splice in an 8-in. section of a bulkhead.
Problem 5. Repair hole 5 × 7 in. in plywood covering on side of fuselage.
Problem 6. Repair hole 3 × 5 in. in top of plywood fuselage covering.
Problem 7. Remove and replace spar of vertical fin.
Problem 8. Replace rib cap strip in rudder rib.
Problem 9. Replace rib cap strip in stabilizer rib.
Problem 10. Replace aileron spar.

INDEX

INDEX

The Aviation Collection by Sportsman's Vintage Press

www.SportsmansVintagePress.com

Aircraft Construction Handbook	by Thomas A. Dickinson
Aircraft Sheet Metal Work	by C. A. LeMaster
The Aircraft Apprentice	by Leslie MacGregor
Aircraft Woodwork	by Col. R. H. Drake
Aircraft Welding	by Col. R. H. Drake
Aircraft Sheet Metal	by Col. R. H. Drake
Aircraft Engines	by Col. R. H. Drake
Aircraft Electrical and Hydraulic Systems, and Aircraft Instruments	by Col. R. H. Drake
Aircraft Engine Maintenance and Service	by Col. R. H. Drake
Aircraft Maintenance and Service	by Col. R. H. Drake

Made in the USA
Monee, IL
14 September 2026

62974508R00118